Charles Henry Stowell

The Students' Manual of Histology

For the Use of Students, Practitioners and Microscopists

Charles Henry Stowell

The Students' Manual of Histology
For the Use of Students, Practitioners and Microscopists

ISBN/EAN: 9783337404604

Printed in Europe, USA, Canada, Australia, Japan

Cover: Foto ©berggeist007 / pixelio.de

More available books at **www.hansebooks.com**

THE STUDENTS'

MANUAL OF HISTOLOGY,

FOR THE USE OF

STUDENTS, PRACTITIONERS AND MICROSCOPISTS.

—BY—

CHAS. H. STOWELL, M. D.,

Assistant Professor of Physiology and Histology, and the Instructor in the Physiological
Laboratory of the University of Michigan.

ILLUSTRATED BY ONE HUNDRED AND NINETY-TWO ENGRAVINGS.

DETROIT:
Geo. S. Davis, Publisher.
1881.

PREFACE.

HISTOLOGY has made such rapid advances within the past few years—not only regarding its place as a part of the science of medicine, but also regarding the new facts discovered by working microscopists of both continents,—that a manual of this character, bringing the subject down to the present time, will, we trust, meet the wants, alike of the student, physician and microscopist.

This manual is not intended to supercede either the large text-book of Stricker or the complete atlas of Klein or other similar works. While these works are necessary and invaluable to the teacher, yet it has been apparent to us for some years that they were altogether both too full and expensive to make them companions of the student; and now that our laboratories are so general that nearly every medical student studies the microscopic structures of the various tissues, even the admirable compend of Frey fails to supply the want, viz., necessary directions for preparing and preserving.

We have endeavored, in this volume, to condense the descriptions as much as possible without injury to its completeness or accuracy.

Of course there are many subjects concerning which our best students and writers materially differ, and as it would far exceed the limits of this work to engage in discussions, we have given in such cases, either those results most generally received or those to which most authority is attached, with, perhaps, the author's own particular views added.

The 'laboratory work' is, by no means, exhausted, nor is it even full. Those methods are given which are most familiar and which have proved the most satisfactory in our hands.

So far as our knowledge goes these are the best methods known, yet others may be as good and it would not be surprising if some were found better.

We have taken some care to discover who was the original owner of the drawings we have taken from other books. We have found nearly everyone of them in several works and no credit given to anyone. When we were not positive, credit was given to to the author of the work in which the drawings were found. Credit should be given to Beale for figure 122. The remaining drawings were all carefully and accurately made from specimens prepared by us, nearly all of which are now in our possession. We believe they can be fully relied upon as correct. The magnifying power used is given in each case, therefore when the size of any object is not given in the text it can be readily ascertained by dividing the size of the figure by the number of diameters it is magnified. The magnifying power was acertained by measuring the micrometer (Roger's) lines at ten inches from the eye-piece.

The subject of the first chapter can be treated but briefly. To obtain anything like a complete knowledge of this subject it will be necessary to consult some of the following works: "Microscopical Technology," Dr. Carl Seiler; "The Microscope and Microscopical Technology," Frey; "How to use the Microscope," Beale.

Our facilities for obtaining tumors have been ample, and their study has taken much of our time.

The illustrations given are from specimens of our own preparing and they convey as accurate an idea as possible of the appearances of these growths as seen under the microscope.

The concluding chapter on the principal starches is introduced because these grains are so frequently encountered in general work, and because the physician or microscopist is so often called upon to examine a specimen with reference to their presence.

I am especially indebted to my former assistant, Dr. D. N. De Tarr (now of the New York State Museum) for most valuable assistance both in the preparing and in the producing on paper of many of the specimens. For the neatness and tact displayed in the production of the book we all are alike grateful to the publisher.

CHAS. H. STOWELL.

"PHYSIOLOGICAL LABORATORY,"
        UNIVERSITY OF MICHIGAN.
March, 1881.

CONTENTS.

5

ILLUSTRATIONS.

CHAPTER I.

The Microscope.

THE word 'microscope' is a compound of two Greek words, μικρός, a small thing, and σκοπέω, to view.

Microscopes may be divided into two general classes, simple and compound. In a simple microscope we look at the object directly, while in a compound microscope we look at the magnified image of the object. Thus the difference is purely an optical one, for a simple microscope may be much more expensive and complex than a compound one, although as a rule the opposite is true.

In the simple microscope the object is seen in its natural position, but in the compound microscope the image is reversed, or inverted. This may be obviated by placing in the body of the microscope a set of lenses termed the erector. Very soon, however, the student becomes familiar with this inversion, and is not annoyed in the least by it.

The compound microscope consists essentially of an object glass, or objective, which magnifies the object, an eye-piece which magnifies the image formed by the objective, a mirror to reflect the light and mechanical appliances.

The Stand of a microscope includes all the framework to which are attached the eye-piece and the objective. Stands are sold separately by many makers, although one or more eye-pieces usually accompany them. The purchaser is thus left free to make his own selection of objectives.

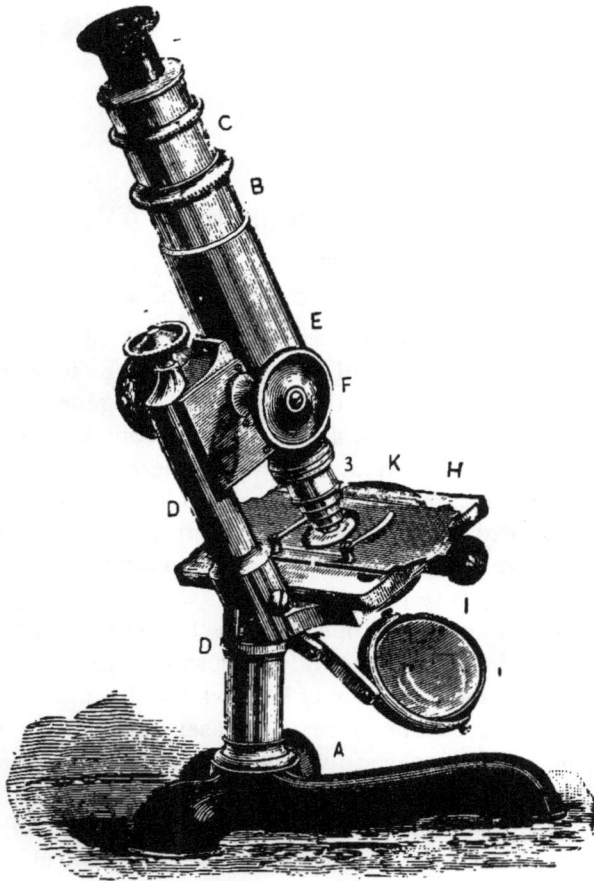

(Cut one-third of actual size.)   [Bausch and Lomb.]

Fig. 1. Compound Microscope. A, the base or foot; B, the body; C, the draw-tube; D, the arm; E, the collar; F, the coarse adjustment; G, the fine adjustment; H, the stage; I, the object-carrier, K, the diaphragm. 1, the mirror; 2, the eye-piece; 3, the objective.

A stand usually consists of the following parts :

The Base or Foot (Fig. 1.) "A." Of all the forms the tripod meets the most general approval.

The Body, "B," that part to which the objective is attached.

The Draw-Tube, "C," which slides within the body.

The Arm, "D," a support for the body. This is usually broken by a joint in order that the instrument may be inclined as seen in the figure.

The Collar, "E," a tube surrounding the body.

The Coarse Adjustment, "F," for coarsely focusing the instrument.

The Fine Adjustment, "G," for more accurate work. This is one of the most desirable things about a stand and should be carefully examined by the purchaser.

The Stage, "H," is that part upon which rests the object to be examined.

An Object-Carrier, "I," is many times combined with the stage in order that the object may be more accurately and carefully moved about. Although not strictly necessary it is a great convenience. "Mechanical" stages are made for that purpose.

The Diaphragm "K," is placed beneath the stage, pierced with different-sized openings, to regulate the amount of light.

Several appliances are sometimes attached to the stand as aids to microscopical manipulation. It would be beyond the limits of this work to enter into their description or to mention the many accessories necessary to complete the outfit.

The Mirror, "1," usually consists of two surfaces, a plane one which reflects the light feebly, and a concave one which concentrates the light upon the specimen. It is attached to a swinging bar beneath the stage in such a manner that light may be reflected from almost any quarter. On some stands it is so arranged that it can be thrown over the stage and the light reflected on the top of an opaque specimen.   This avoids the necessity of an extra condenser.

The eye-piece, "2" consists of two glasses mounted in either hard rubber or brass.   Midway between them is a diaphragm to cut off the outer rays of light.   The eye-piece in most general use is known as the negative or Hughenian. In this eye-piece the convex side of the lenses is directed downward.   The lens nearest the eye for this reason is called the eye-glass, and the one farthest from the eye, and nearest the field, is called the field-glass.

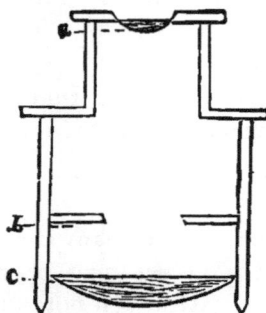

FIG. 2.   Eye-piece in section.   a, eye-glass; b, diaphragm; c, field-glass.

The magnifying power of eye-pieces is designated by either numbers or letters.   In this country letters are chiefly employed.   The lowest power is known as "A" or No. 1; higher powers are known as "B" or No. 2, "C" or No. 3, and so on.   The greater the magnifying power, the shorter will be the eye-piece.   The short eye-piece, or the one with high power, is also known as the deep eye-piece; the longer, or the one with less power, as the shallow eye-piece.   One eye-piece, then, may be "A" or low, or shallow; another may be "D" or high, or deep.

As an eye-piece does not magnify the object itself but the image of the object produced by the objective, it will be seen how any imperfection in the objective will be augmented. High eye-pieces should be used only with fine first-class objectives.

The objective, "3," is usually composed of one or more systems of glasses. A system consists of two or more glasses. It is not made of a single glass because the powers of refraction and dispersion are not equally united in any single refracting medium. That is, in the same power of refraction one medium may give a much greater deviation to the colored rays than another.

Crown and flint glass act with regard to each other in such a manner that if a crown glass lens be united with a flint glass lens, the refraction of the former is lessened by the dispersive action of the latter, while the color dispersion of the former is neutralized by the opposite action of the latter. Spherical aberration may be largely remedied by this same combination. The lenses are firmly cemented together by Canada balsam or Dammar. The glasses thus united constitute a system, and in Fig. " 3," three of these

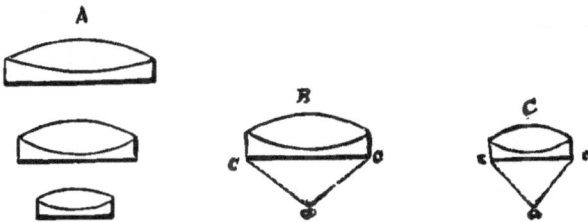

FIG. 3. A. Achromatic objective of three systems; B, objective with high angle of aperture C, objective with low angle of aperture. The angle of aperture is the angle c a c.

systems complete the objective. These systems are mounted in either brass or hard rubber, which at the upper end is provided with a screw of standard size. Such a sized screw is called a "Society Screw" and will fit in the body of any first-class stand.

An objective should possess the following good qualities :

1. Defining power.
2. Resolving power.
3. Freedom from spherical aberration.
4. Penetrating power.
5. Freedom from chromatic aberration.
6. Flatness of field.
7. Working distance.

Defining power is without question the most important quality to be sought in a lens.   Its presence makes the objective of the utmost value, and its absence renders it simply worthless.

Defining power gives a clear, distinct and sharply cut outline.   Its absence is denoted by haziness, indistinctness and want of clearness.

Resolving power enables closely approximated markings to be seen distinctly.   While defining power shows the outline of a specimen well, resolving power enables the observer to detect the most intricate structure on its surface.

Spherical aberration exists when the peripheral and central rays do not actually reunite in a point.   Those rays passing near the periphery, being more strongly refracted, come to a focus sooner than those which pass through the central portion.   Now if some parts of a lens bring the rays to a focus sooner than other parts, they must magnify more, thereby distorting the figure.   This is found to be the case with all objectives having spherical aberration, causing what is known as " aberration of form."

With penetrating power we look deep into the structure of the object.

Chromatic aberration exists when a ray of light is not refracted as a whole, but is decomposed into rays of various colors, which are refracted in different degrees, forming a spectrum.   All objects examined now are seen fringed with colors

An objective is said to be "achromatic" when it is nearly, if not quite, free from this aberration.

It is impossible to perfectly remedy these two aberrations, but by the use of the two kinds of glass mentioned above, they are nearly obviated.   Objectives thus made are said to be " corrected."

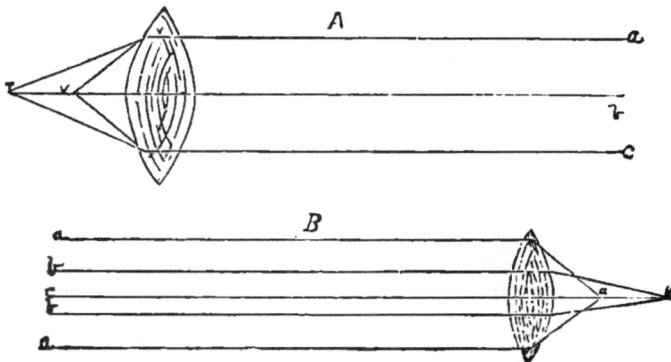

FIG. 4.   A, chromatic aberration ;   a, c, rays of white light; v, violet rays ; r, red rays. B, spherical aberration.

A Field includes all that is presented to the eye through the microscope.   It is said to be flat when all parts of it are in focus at the same time. An objective with good defining power is very liable not to have a flat field, and a perfectly flat field is usually associated with poor defining power.   Defining power should never be sacrificed for flatness of field, since with a good, fine adjustment the latter is easily remedied, while nothing can restore the former.

Working distance is the distance between the front glass of the objective and the point in focus.   Some manufacturers make objectives with a large working distance without materially affecting their defining or magnifying power.   For many purposes such objectives are of great value.   For a discussion of the vexacious question of "angular aperture" we refer our readers to works on microscopical technology or to

many articles in the various microscopical journals.   The
usual definition is this :

Angle of aperture is the angle formed by two lines extending
from the point in focus to the opposite sides of the aperture of
the objective.   While one school claims that a high angle has
great resolving and poor penetrating powers, another school
earnestly urges that both can be combined in the same objective.

Immersion objectives are those that require a drop of
liquid between the end of the objective and the cover glass.
Water is generally employed for this purpose, although glycerine, on account of its greater density, is sometimes used.
By employing a liquid in this way the glass surfaces of the objective and cover glass are, to a certain extent, extinguished
and thus a considerable loss of light prevented ; at the same
time the refraction of the rays of light at the upper surface
of the cover glass is very much diminished, so that many more
rays of light pass into the microscope.   The specimen is then
better illuminated, and also better defined.

Objectives are numbered according to their magnifying
power.   In this country the system is different from that
abroad, where they are numbered 1, 2, 3, 4, 5, etc.   Here an
objective is known as an inch, one-half inch, one-fourth, one-
eighth, one-thirtieth, etc.   These terms refer solely to the magnifying power.   For instance, a one-fourth inch objective has
the same magnifying power as a single lens whose focal distance
is one-fourth of an inch.

Each microscope should have with it a micrometer.

Nothing can be more convenient or useful than a good
eye-piece micrometer.   Knowing the value of the spaces
to which it is ruled, objects can be accurately and quickly
measured.   Of course, the value of these spaces will depend upon
the objective used and the length of the tube of the microscope.   By always using the draw tube fully extended the length
of the tube will be fixed, and with the aid of a stage micrometer

ruled 100, 1000, 2000 lines to the inch, the value of the spaces of the eye-piece micrometer for the various objectives is reckoned once for all. This is done in the following manner : Bring in the field the lines of the stage micrometer $\frac{1}{1000}$ of an

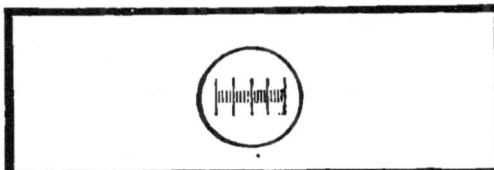

FIG. 5. Eye-Piece Micrometer. (Increased one-third.)

inch apart ; place the eye-piece micrometer in its proper place in the eye-piece ; notice how many spaces of the eye-piece micro-meter cover one space on the stage micrometer. Using the ¼ inch objective and the " C " eye-piece, we will assume that five spaces of the eye-piece micrometer cover one on the stage mi-crometer. Then one space on the eye-piece micrometer re-presents the $\frac{1}{5000}$ of an inch. Now remove the stage micro-meter and place in the field a specimen of blood, for instance, a white blood corpuscle is seen just to fill two spaces in the micrometer. It is then the $\frac{2}{5000}$ of an inch in diameter.

To determine the magnifying power of a microscope, it is inclined until the eye-piece is ten inches from the table. The lines of a stage micrometer are then accurately focused. By means of a "Camera Lucida" or "Neutral tint glass reflector" (Fig. 6), the magnified image is thrown upon a sheet of paper resting on the table and directly beneath the eye-piece. The lines are traced with a pencil while the eye is in the position noted in the figure, and their distance apart measured with a scale. This distance is divided by the distance between the lines on the stage micrometer, and the result will be the number of diameters the instrument magnifies—not the num-ber of times or areas, which would be the square of the diameters. Or the following method : Place a scale in front

of and ten inches below the eye-piece.   By looking in the
instrument and keeping both eyes open, the lines of the stage
micrometer can be seen resting on the scale, when their
distance apart can be noted.   Divide this distance by that
between the lines on the micrometer, and the number of
diameters will be given.

Having no eye-piece micrometer, the size of any object is
obtained in the following way :   Assuming our microscope to

FIG. 6.   Camera Lucida, or Neutral Tint Reflector.   (Bausch and Lomb.)

magnify 500 diameters, the specimen to be measured is substituted for the stage micrometer, and its image thrown down
on the paper as were the lines of the micrometer, and its size
measured with a scale.   This measure is divided by the magnifying power of the instrument.   Thus a red blood corpuscle
appears on the paper $\frac{1}{8}$ of an inch in diameter.   It has been
magnified 500 diameters. Its true size then is $\frac{1}{500}$ of its
apparent size, viz. $\frac{1}{4000}$ of an inch.

A microscope is said to be " in focus " when the specimen is seen to the best advantage. For the higher powers the following rule should be observed : Incline the head until the eye is on a level with the stage. With the coarse adjustment place the objective very near the cover glass, within its focal length. Then, while looking in the microscope, focus up. If this rule be carefully observed, the breaking of cover glasses and the destruction of specimens will be materially diminished.

For general microscopical work daylight is to be preferred. Not strong direct sunlight, which is only useful under special circumstances, but such an even, steady light as can be found by a window looking to the north. Nothing can take the place of this northern light, both when the sky is clear, and when, best of all, the sunlight is reflected from a white cloud. While gas-light and lamp-light are inferior to daylight and weakening to the eyes, direct sunlight is positively injurious.

Transparent objects may be viewed by either direct or reflected light.

When the light passes directly through the specimen and microscope without having been reflected by the mirror, it is said to be direct.

If the mirror be so placed that the reflected rays are in the optical axis of the microscope, the light is said to be central.

If the mirror be turned to one side so that the rays pass through the object at an acute angle, oblique light is obtained.

In the care of the microscope the following practical hints may not be out of place :

When removing from, or placing on the stage a specimen, if the higher powers have been used, always raise the body of the instrument.

It is rarely necessary to clean a good microscope.

Use soft. chamois to clean, and camel's hair brushes to dust.

Remove balsam, etc., from objectives by slightly moistening the chamois in turpentine and carefully wiping it off.

Avoid handling the instrument.

Carry it by the arm.

Always clean immersion objectives thoroughly, and immediately after using.

When not in use, keep the instrument in its case or under a bell jar.

However ; better let the instrument wear out rather than rust out.

For practical work a good microscope need have but two eye-pieces of different powers, and a 1 in. and ¼ in., or ¾ in. and ½ in. objectives.

FIG. 7. Turn-Table. (R. and J. Beck.)

The beginner will need a pair of fine dissecting forceps, curved scissors, a knive or two, a few needles, a razor flat on one side and concave on the other for making sections, a few camel's hair brushes, chamois skin, glass slides and cover glasses. These, with the following list of twelve reagents, will complete the necessary outfit. Other reagents and instruments will be added as their need becomes manifest. Many of them, however, the ingenuity of the worker, who is weaker in his pocket than in his head, will extemporize. Notably a "turn table" and "microtome." (Fig. 7 and 8.)

LIST OF REAGENTS.

1. Normal saline solution, ¾ p. c. solution.
2. Glycerine.
3. Alcohol.
4. Ether.
5. Acetic acid.
6. Iodine solution.
7. Canada balsam.
8. Carmine staining.
9. Hæmatoxylin.
10. Oil of cloves.
11. Dammar.
12. White zinc cement.

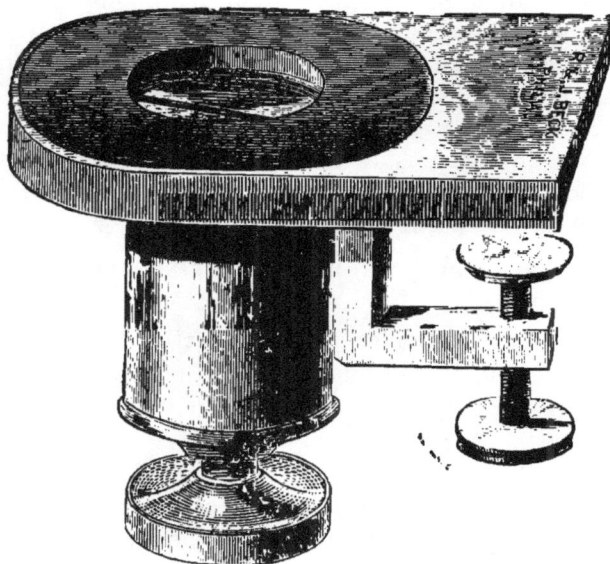

FIG. 8. Microtome. (R. and J. Beck.)

For anything like a complete list of the various reagents, injecting and staining mixtures, and the methods of preparing them, the reader is referred to special works on those subjects. We append a few formulæ in general use.

BEALE'S PRUSSIAN BLUE, FOR TRANSPARENT INJECTIONS.

Common glycerine, - - - - - 1 ounce,
Spirits of wine, - - - - - - 1 ounce,

Ferrocyanide of potassium,   -   - 12 grains.
Tincture perchloride iron,   -   - 1 drachm.
Water, - - - - - - - - - 4 ounces.

The ferrocyanide of potassium is to be dissolved in one ounce of the water and glycerine, and the tincture of iron (muriated tincture of iron) added to another ounce. These solutions should be mixed together very gradually and well shaken in a bottle. The iron being added to the solution of the ferrocyanide of potassium. When thoroughly mixed, the solutions should produce a dark blue mixture, in which no precipitate or flocculi are observable. Next the spirit and the water are to be added very gradually, the mixture being constantly shaken in a large stoppered bottle. In cases, in which a very fine injection is to be made for examination with the highest powers, half the quantity of iron and ferrocyanide of potassium may be used.

BEALE'S ACID CARMINE INJECTING FLUID.

Carmine, - - - - - - - - - - - 5 grains.
Glycerine, with 8 or 10 drops of acetic or
          hydrochloric acid, - - - - - ½ ounce.
Glycerine, - - - - - - - - - - - 1 ounce.
Alcohol - - - - - - - - - - - 2 drachms.
Ammonia, - - - - - - - - - - - a few drops.

Mix the carmine with a few drops of water and, when well incorporated, add about five drops of liquor ammoniæ. To this dark-red solution about half an ounce of the glycerine is to be added, and the whole well shaken in a bottle. Next, very gradually pour in the acid glycerine, frequently shaking the bottle during admixture. Test the mixture with blue litmus paper, and if not of a very decidedly acid reaction, a few drops more of acid may be added to the remainder of the glycerine, and mixed as before. Lastly, mix the alcohol and water very gradually, shaking the bottle thoroughly after adding each successive portion till the whole is mixed.

STAINING MIXTURES. BEALE'S STAINING CARMINE, FOR STAIN-
ING GERMINAL MATTER.

Carmine, - - - - - - - - 10 grains.
Strong liquor ammoniæ, - - ½ drachm.
Price's glycerine, - - - - - 2 ounces.
Distilled water, - - - - - - 2 ounces.
Alcohol, - - - - - - - - ½ ounce.

The carmine in small fragments is to be placed in a test
tube, and the ammonia added to it. By agitation and with
the aid of the heat of a spirit lamp, the carmine is soon dissolved. The ammoniacal solution is to be boiled for a few
seconds, and then allowed to cool. After the lapse of an hour,
much of the excess of ammonia will have escaped. The glycerine and water may then be added and the whole passed
through a filter or allowed to stand for some time and the perfectly clear supernatant fluid poured off and kept for use.

HÆMATOXYLIN.

Make a saturated solution of crystallized calcium chloride
in 70 per cent. alcohol. Shake and let stand. Add alum to
excess. Shake well, let stand, and then filter. Make a saturated solution of alum in 70 per cent. alcohol. Add this to
the above filtrate in the proportion of 8 to 1. To this mixture
add drop by drop a saturated solution of hæmatoxylin in absolute alcohol until it has a somewhat dark purple color.
Too deep staining can be removed by placing the section in
dilute acetic acid.

ANILINE BLUE-BLACK.

Dissolve 5 grains of aniline blue-black in 100 c. c. of water. Dilute with water to any strength required.

The following are useful for

MOUNTING MEDIA.

Canada Balsam.
Canada balsam to be used without heat. Prepared as
follows : Heat some of the balsam over a sand bath until it is

hard when cold. Then dissolve it in a small quantity of benzole. To mount in Canada balsam the specimen is thoroughly
saturated with alcohol. The excess of this removed with
strips of cut blotting paper and oil of turpentine added. As
soon as the section has become saturated or cleared the excess
of oil is removed and the balsam added.

Dammar. This is a great favorite with most histologists.
It renders the tissues more transparent than balsam and is a
convenient fluid to handle. It is prepared as follows : One
half ounce each of dammar resin and gum mastic is dissolved
in 3 ounces benzole and filtered. To mount a section in dammar, it is first left in alcohol, than transferred to absolute alcohol until no water is in the section. The excess is removed
by blotting paper and the oil of cloves added. Here it is allowed to remain until transparent. If it does not clear in a
short time, in all probability the alcohol did not entirely remove the water. Alcohol should be added again and allowed
to remain longer on the specimen. Adding the oil a second
time will doubtless clear up the section completely. Then remove the excess of oil with blotting paper and add a drop or
two of dammar and cover with thin glass.

If glycerine or other fluid mounting medium be used, it
will be necessary to make a cell in which can be placed the
fluid and specimen. Cells are made with either of the following cements : Gold size, Brunswick black, White zinc. The
border may be oval, square or circular,—if circular, a turn-
table is employed in order that the circle may be true and sym-
metrical. This soon hardens and forms a firm support for the
cover glass, the edge of which should come just to the centre
of the border. An extra layer of cement is now added, one-
half of which reaches on the cover glass and the other half on
the glass slide.

For embedding mixtures the following are especially recommended :

Solid paraffin,	3 parts	
Cocoa butter,	1 part	soft.
Hog's lard,	3 parts	

Solid paraffin,	3 parts	
Cocoa butter,	2 parts	hard.
Spermaceti,	1 part	

Solid paraffin,	2 parts	
Cocoa butter,	1 part	harder
Spermaceti,	1 part	

Paraffin,	2 parts	transp't and
Vaseline,	1 part	easy to cut.

FIG. 9. Injecting Apparatus.

For further information on these subjects the reader is referred to Beale on "How to Work with the Microscope," "The Microscope in Medicine," or to the admirable work of Frey's, "The Microscope and Microscopical Technology."

Figure 9 illustrates a cheap injecting apparatus. *a* represents a pail partly filled with water, which can be raised or
lowered, to regulate the pressure, by fastening one end of a
cord to the handle of the pail and then passing the other end
over a pulley fastened to the ceiling of the room ; *b* is a bottle
with an air-tight fitting cork, pierced by two short glass tubes ;
c is a bottle partly filled with the injecting mixture. Through
the cork of this bottle are two glass tubes, one of which is
short, while the other reaches very nearly to the bottom of the
bottle , *d* is a brass nozzle with a stop-cock ; *r,* is rubber tubing, which unites the different parts as seen in the figure. A
y-shaped glass tube can be inserted midway in the rubber tube
between the two bottles, so that two bottles of the injecting
mixture can be attached to the one large bottle, *b,* which is
empty at first. A third glass tube can be placed in the cork
of the bottle *c,* which can be united by rubber tubing to a U
shaped glass tube partly filled with mercury, and thus the
amount of pressure obtained. By raising the pail, the water
descends the rubber tubing and compresses the air in the
bottle *b.* The air is forced through the middle piece of rubber
tubing and presses on the top of the injecting mixture in the
bottle *c,* which is forced up the glass tube, along the rubber
tube to the canula, and into the animal or organ to be injected.

CHAPTER II.

The Amœba and the Cell.

LOW down in the scale of animal life are found minute organisms of variable size inhabiting stagnant water, mud, and water in which animal matter has been infused and exposed to the direct rays of the sun. They have the appearance of a particle of the white of egg, clear and transparent, perhaps slightly granular, quite fluid in the centre and of firmer consistency towards the periphery.

They are especially remarkable for their incessant and rapid changes of form, causing them to move about, but not in any particular direction. Their movements are effected by a flow of their protoplasm, causing them to thrust out prolongations, known as pseudopodia. The dense exterior we know as the ectosarc, the more granular fluid interior, the entosarc. In some amœbæ there appears a clear spot which dilates to a certain extent, then contracts rapidly and disappears, to reappear again with tolerable regularity. This is the contractile vesicle. It seems to serve two purposes, first, as a pump to force water into and out of the body, and second, as a means of procuring food, for when dilated to its full extent it will sometimes contract with such vigor as to break through the ectosarc and cause its contents to rush out into the liquid in which it lies. It then dilates, causing a strong suction force, which draws in a certain amount of water and with the water, infusoria, entozoa and vegetal forms, the food of our amœba.

A nucleus and nucleolus are occasionally seen.

29

The first thing noticed in examining one of these little animals is, it is contractile. Its peculiar amœboid movements, its flow of protoplasm, are identical in their fundamental nature with the movements occurring in a muscle during its contraction. The second is, it is irritable and automatic. If a foreign body be brought in contact with an amœba when it is at rest, movements result. These movements are not passive in their nature, proportionate to the force employed, but are the result of an explosion of the energy of its living matter. Rarely does one see the amœba at rest. It is almost constantly changing its form, not from external stimuli, but from changes of its substance, the cause of which lies within the body itself. The marked features of nervous tissue are its irritability and automatism. The third is, it is secretory and excretory. Besides the method described above it will be noticed that our amœba

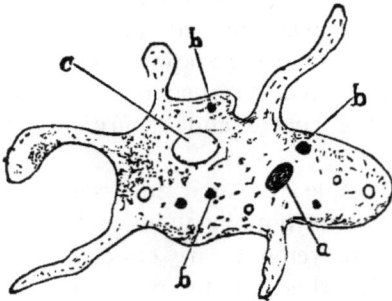

FIG. 10. Amœba. a, nucleus; b, foreign bodies; c, vacuole.

has another way of procuring food,—by extending around it its pseudopodia until the particle is completely surrounded by the living matter. Here the foreign body remains for a time. If it be suitable for food it soon becomes changed into material like the mass surrounding it, — into "amœba stuff." Part of it may be changed and the remainder thrown off as excrementitious matter or the whole may be served in a similar manner. If all the particle be not assimilated then the amœba simply moves away by its flow of protoplasm and leaves it behind. There must be chemical products present for the purpose of dissolving and effecting changes in this raw new material taken as food. These must be regarded as secretions. Our amœba is certainly excretory. In man, the digestive, urinary, and pulmonary tracts, and the epithelia represent this physiological property of the amœba.

The fourth is, it is metabolic. Constantly undergoing chemical change. Certain cells in the human body are specially reserved for carrying on chemical changes. Their material is derived from the blood and their products are finally returned to it. Such cells are the fat cells, liver cells, also the lymphatic and ductless glands, and in one sense, the blood.

The fifth is, it is reproductive. After attaining a certain size or living a longer or shorter life, it may by division resolve itself into two parts, each of which is capable of living as a complete unit. The amœba divides by becoming constricted in its centre, by its protoplasm flowing in opposite directions, or by a pseudopodium detaching itself from the body of the cell. Certain cells are set apart in the human body for the accomplishment of this purpose. Such collections of cells are the ovary and the testis. Man, then, is but a federation of amœbiform units. Certain of these units have been exclusively set apart for the manifestation of certain of the properties of protoplasmic matter. These groups have received the name of "tissues." With this grouping there has come a change in structure in order that the part might better perform its function. At one period in the history of these cells they were as simple as our amœboid unit, in fact, for that matter, at an early period in the history of every life the whole being, the embryo, was but a mass of units as simple in their

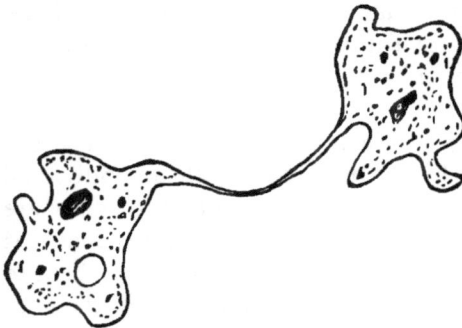
FIG. 11. Amœba dividing.

structure as the amœba. Some cells remain in this amœboid
condition in the body for a considerable time ; such cells are
the mucous and pus corpuscles, and the white corpuscles of
the blood.

At a bound, then, we pass from this low creature to the
highest to find that the tissues of the latter are but collections
of the former. So a tissue is chosen and with the aid of the
knife and needles its parts are for a time successfully separated.

But at last a period arrives when even this will not
answer, and we turn to the microscope to find our tissue
infinitely compounded of thousands of the smallest elements.
To discover and to examine these constitute the science of
tissues, or histology.

While the cell existed as an amœba it acted in an independent manner, but now that it is in the service of a unity of
cells it is a subordinate and must conform itself to its sur-
roundings.

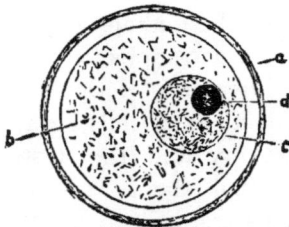

Each cell in the body then is a
living individual with an individual
function. Some of these cells are
very small, for we shall see that it is
possible for five millions of them to
be contained in a particle of the substance of the body no larger than a
cubic millimetre. While some are
so large they are nearly, if not quite,
visible to the unaided eye. In shape,
also, there is the greatest variation. First, there is the
spheroidal cell, from which the bodies of all the higher
animal have proceeded, the ovum. As a result of com-
pression and adaptation come other cells, from the slender
cylindrical to the flat scaly. Still other cells appear with
branched processes growing from their bodies in opposite directions. There exists, then, every variety of shape

FIG. :2. Human ovum.   a, vitelline
membrane ; b, vitellus; c, germinal
vesicle ; d, germinal spot.

and form in the fully developed cells, although we shall see further on that early in their history they were all alike, simple undifferentiated bioplasm. In a well developed epithelial cell from the surface of the tongue two parts are readily recog-

nized. First, the nucleus, a round or oval body, occupying a small part of the cell near or in its centre, and, second, the part of the cell surrounding this.

In the alkaline solution of carmine (page 25) we possess an agent capable of coloring the different parts of a tissue or

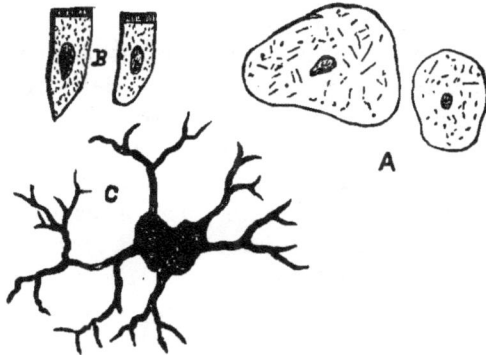

FIG. 13.   A, flattened epithelial cells; B, cylindrical cells; C, branched connective tissue cell. x 400.

cell to different degrees. In a cell it is noticed that the innermost part is invariably colored the most intensely. In the case of the epithelial cell the nucleus will be stained a deep red by the carmine, while the outer part remains unaffected. Now if any young or rapidly growing epithelial surface be examined, as the layer of epithelium over a papilla of the tongue, those cells nearest the blood-vessels will be seen to take the carmine staining completely. They are composed of nucleus matter alone; from this matter must come all the future parts of the cell. No matter how high

FIG. 14.   Fully developed cell. a, formed material; b, nucleus; c, nucleolus.

or complex the tissue, it must proceed from this first living germinal matter, the bioplasm of Beale. If the cells be examined a little further from our nutrient vessel a material will be seen sur-

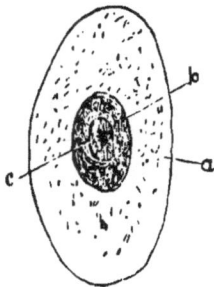

rounding the nucleus, which does not take carmine staining. This was once living nucleus matter but now from coming in contact with air or fluids death occurs upon its surface and the nucleus or germinal matter becomes changed to lifeless formed material. Still farther away on the surface of the papilla the nucleus has nearly disappeared. It is now so far removed from its supply of pabulum that it has become gradually changed. Nutritious material then is deposited from the blood, first, in the centre of the living part of each cell, and while the inner part of each cell, the nucleus, is being constantly replenished, its outer part is as constantly passing into lifeless formed material. All matter must be nucleus matter before it can become formed material. Only nucleus matter can be said to live. It lives, because it is capable of converting material unlike itself into material like itself. The nuclei of muscle convert common pabulum into muscle nuclei which is thence converted into muscle formed material. The nuclei of nerve cells are capable of taking some of the same pabulum and converting it into nerve nuclei and thence into nerve formed material. Thus, man, a federation of these cells is capable of converting his food into materials like his own body, hair, nails, skin, etc., while the dog by eating of the same food will convert it into its own peculiar tissues. The youngest, most recently deposited matter of a tissue is found in its nuclei, the oldest in its formed material. There is a law in the body by which the amount of pabulum supplied to the nuclei by the blood, just equals the wear and waste of the cells.

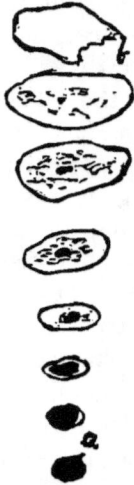

FIG. 15. Illustrating diminishing nuclei in cells as they approach the surface. a, cells near blood-vessels; b, cells remote from blood-vessels.

But if from any cause the part be irritated, be spurred to in-creased action, then an extra amount of blood flows to the part, an extra amount of pabulum is furnished to the germinal matter, causing it to increase rapidly.   Now from two to five nuclei are seen in one cell. If the process goes on the changes become so rapid that the germinal matter does not change into formed material, and there now appear a multi-tude of round, globular bodies, in cells, and on the surface of cells, familiarly known as pus.   Doubtless some of these cells are the migrated white corpuscles of the blood, but the mass of them represent living germinal matter undergoing rapid changes and possessed of a low vitality.   This germinal matter may come from any irritated cell.   It must be borne in mind that the supply, and change of pabu-lum to the germinal matter, and from it to the formed material of the body, are constant and uninterrupted.   From what has been said it must be evident that com-pared to human life, the life of any one cell is very short.

FIG. 16.   Pus in epithelial cells found in urine. x 250.

When we consider the immense numbers removed daily from the surface of the body by the friction of clothing— laying aside the work of the sponge and towel—some idea can be gained how active must be the changes going on just be-neath the surface.   Add to this the number rubbed off by every act of moving the tongue, in speaking, drinking, and eating, and we commence to understand how most of the cells are destined to an early death.   This great loss is replaced by the formation of new cells and by the division of those already formed,—the nucleus dividing first, then the whole cell becoming separated into two by constriction.   In order to understand the most complex we study the most simple.   That we may the better understand the highest, we watch the lowest, for they all receive new material and transform it into the constituents of their own bodies. They live, they grow, they reproduce their kind, they die.

Our knowledge of the structure of cells and of their nuclei has been greatly increased by the labors of such men as Kleinenberg, Heitzman, Auerbach, Flemming, Klein and others.

In No. 71, 1878, p. 315, and in No. 74, 1879, p. 125, of the Quarterly Journal of Microscopical Science, are articles by E. Klein, giving personal observations on the structure of cells and nuclei. The first article opens with a resumè of the work of several observers. From this we learn that Frommann described a network of fibrils in the nuclei of many kinds of cells as early as 1867.

In 1873, Heitzmann asserted that the substance of various cells, amœbæ, blood-corpuscles, cartilage cells, bone cells, epithelial cells, etc., contains a network of minute fibrils, into which pass fibrils radiating from the interior of the nuclei of those cells.

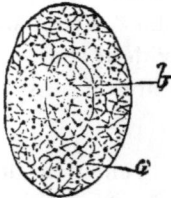
FIG. 17. Epithelial cell. a, intracellular network; b, intranuclear network. (Klein)

To demonstrate this structure Klein places the tissue into a 5 per cent. solution of chromate of ammonia in a closed vessel. It is kept here for about 24 hours. It is then washed in water for about half an hour, when it is placed in a dilute solution of picro-carmine, where it is left till it assumes a deep pinkish-yellow tint.

Examined in glycerine the nuclei show a beautiful network of fibrils, "Intranuclear network." The nucleus is surrounded by a membrane which is in connection with the fibrils of the nucleus.

The fibrils vary in the different nuclei. They may be fine, delicate and smooth, or coarse, of irregular outline and convoluted. Klein regards the "granules," or minute bright spots seen in nuclei, as representing the fibrils of the net-work seen in optical transverse section or at the point of anastomosis. Some fibrils, however, are possessed of irregular thickenings. He believes the so-called "nucleoli" are accumulations or local thickenings of the fibres of the intranuclear network. In many cells there exists a delicate fibrillar network in the cell substance—in the "formed material" of Beale. "Intracellular network." There is a direct, anatomical continuity of the fibrils of the intracellular network with those of the intranuclear network. In the meshes of this fibrillar network is a homogeneous "interfibrillar substance."

CHAPTER III.

Blood.

BLOOD may be described as a tissue, the discs forming the essential element, the cells ; while the plasma represents the matrix in a liquid condition. If the matrix of bone could be liquefied so that the bone cells could freely move about, it could be well compared to blood as a tissue. Or if some re-agent could be applied to completely solidify the plasma, we should have a tissue not very unlike cartilage.

As early as 1661 Malpighi discovered little particles in the blood of the hedgehog, which he thought were little particles of fat, but these were really the corpuscles of the blood. Only a few years later Leuwenhoek described quite minutely the corpuscles, which he had discovered in human blood, although the very best lenses at his disposal were made by himself and did not magnify over 160 diameters. To him is given the honor of first discovering the corpuscles of the blood. It was not until a century later that another kind of corpuscles was discovered by Hewson, known as the white or colorless corpuscles, white globules, or the "leucocytes of Robin." Beside the liquid plasma, blood consists of,—

First, red corpuscles.
Second, white corpuscles.
Third, accidental or temporary ingredients.

The corpuscles and plasma bear the following relations to each other.

	BY VOLUME.	BY WEIGHT.	SPECIFIC GRAVITY.
Plasma,	60 per ct.	55 per ct.	1030 ⎫
Corpuscles,	40 per ct.	45 per ct.	1088 ⎭ 1055

The blood is distributed through the body of man in the following proportions :

One-fourth in the heart, lungs, large arteries and veins.
"   "   "   "   liver.
"   "   "   "   skeletal muscles.
"   "   "   "   other organs.

In the organs of the rabbit Ranke found the blood distributed as follows :

	PER CENT. OF TOTAL BLOOD.
In the spleen, - - - -	.23
" " brain and spinal cord,	1.24
" " kidneys, - - - -	1.63
" " skin, - - - - -	2.10
" " intestines, - - - -	6.30
" " bones, etc., - - -	8.24
" ' heart, lungs and great blood vessels, - -	22.76
" " skeletal muscles, -	29.22
" " liver, - - - - -	29.30

The total quantity of blood in the human body is estimated by different writers at from $\frac{1}{8}$ to $\frac{1}{13}$ of the body weight.

In the rabbit, $\frac{1}{18}$ of body weight.
"   "   dog, $\frac{1}{15}$ "   "   "
"   "   cat, $\frac{1}{21}$ "   "   "
"   "   frog, $\frac{1}{15}$ "   "   "

THE RED CORPUSCLE.

As seen in a single layer under the microscope the red corpuscles are of a yellowish-green tint, and it is only when seen in masses, that they present a reddish color.   These red corpuscles are found in the blood of all the vertebrates, even in the lowest form, the amphioxus.

THEIR ORIGIN AND DEATH.

They are present in the blood of the embryo when the fœtus is little more than $\frac{1}{10}$ inch in length.   At this time they are much larger than those found in the adult, varying in size from $\frac{1}{600}$ to $\frac{1}{1600}$ of an inch in transverse diameter.   In shape they are circular, oval, or globular.   Nearly all have a nucleus readily seen without the aid of reagents.

What is the origin of these primary red corpuscles of the embryo?

Early in the history of the embryo the rudimentary heart consists of a mass of epithelial cells, and radiating from it are two or more tracts—generally one on each side—which, by their subdivision, form the vascular area.

These cells are nucleated and vary in shape according to the pressure to which they have been subjected.   In size they agree with the early red corpuscles described above.

At a certain time some of these nucleated cells in the interior of the mass composing the rudimentary heart, become loosened from their fellows.   The exact time of this occurrence and its cause are not known.   There are certain normal functions of the body performed in a regular way, the cause or causes of the regularity remaining in obscurity.   We only know that these particular cells are separated from the rest to serve a special purpose as carriers of oxygen.

The remaining cells become transformed into the tissue composing the walls of the vessel, which, by twisting upon itself, finally becomes the heart.   There is reason to believe that throughout the vascular area, cells in the interior of the blood-tracts become loosened from their fellows, while the remaining ones are metamorphosed into the walls of the vessels. These loosened cells may be either slightly or quite deeply colored. It would seem that the hæmoglobin is deposited as small granules in different parts of the cells, to become evenly disseminated afterwards.

At this time there are large, circular, oval, nucleated red blood-corpuscles, identical with those seen as late as the middle period of uterine life. They increase greatly by cell division, at least until the embryo reaches a certain age, after which their multiplication may be due to other causes.

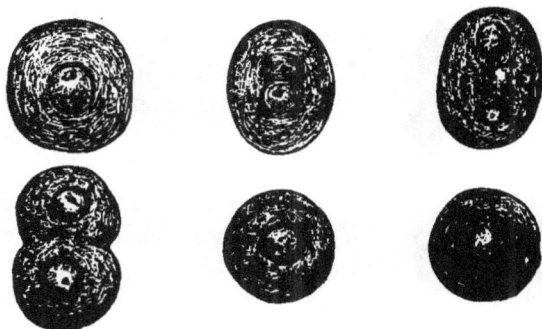

FIG. 18. Red blood-corpuscles of the human embryo, undergoing cell-division.
(From Kirke.)

The development of the red corpuscles in the adult is, and must be, different from their embryonic origin. The basis upon which this assertion rests must be stated, for it might be said that the corpuscles in the adult are either the identical ones found in the embryo, or that they are formed from these by cell division.

The first statement cannot be true, for there is every reason to believe that the red blood-corpuscle is exceedingly short-lived, (see Foster's *Physiology*, 3d edition, p. 35.) The number of corpuscles in the blood varies greatly at different times, as is proved by counting them. Again, after hemorrhage or disease, the normal amount may be regained in a very short time. If the urinary and bile pigments are derived from the hæmoglobin, the number of red corpuscles destroyed must be very great. The second assertion cannot be true, for the corpuscles very seldom, if ever, increase by cell division in the adult (*Ibid.*, p. 36.)

They must, therefore, have an origin entirely distinct from that of the embryonic cells.

The following serves to strengthen an old theory and answer some objections to it.

If we take the pulp of the spleen and examine it carefully, there may be seen large circular cells, colored with hæmo-globin. These cells are, perhaps, the protoplasmic cells of Kölliker. Some of them contain in their interior, the remains of from one to ten red corpuscles. The reason why these very large cells are not found in the circulation, is probably because they are too large to enter the venous capillaries (see histology of the spleen.) Their large size is attained by appropriating to themselves, through their amœboid movements, the remains of one or more red corpuscles ; this operation must take place in the spleen pulp outside of the vessels. Their size will prevent them from entering the first venous capillaries, until they have undergone cell division. This division may be due to the same cause that keeps the amœba about an average size, *viz.*: the attraction of its constituent particles for each other not being equal to the external pressure after they attain a certain growth. As a result of cell division, spherical, nucleated, colored corpuscles

FIG 19.   Blood cells from spleen pulp
(Harley & Brown.)

would be produced, sufficiently reduced in size to enter the circulation ; and we have proof positive that they do enter as suggested above. Special precautions must be taken to demonstrate the presence of these corpuscles in the circulating blood, and Schmidt believes them to be always present in normal blood, in limited numbers. They are also seen in the medulla of bone. They have the appearance of white blood-corpuscles, colored with the hæmoglobin of the red.

When lymph, taken from the thoracic duct or any other lymph vessel in the system, is examined immediately, it is found to be colorless, or nearly so ; but when allowed to clot, it assumes a decided pinkish tinge, which, by microscopical examination, is found to be caused by the presence of red blood-corpuscles. The red corpuscles appearing so constantly after the withdrawal of the lymph from the body, could hardly have an accidental origin. (Dalton's *Physiology*, 6th ed., p. 368.)

Recklinghausen saw the white cells of frog's blood develop into red corpuscles, even when out of the body. (*Arch. für Mic. Anat.*, 1866, p. 137.) Were there not such a difference between them in structure and form, these facts would lead to the conclusion that the white corpuscles give origin to the red. Kölliker, Neumann, and Schmidt are of the opinion that the nucleus disappears from the white cells, while Huxley holds that the red corpuscles represent the bare nucleus of the former. Beale has taught us, that as the cell grows in age its nucleus diminishes in size. His method of staining certainly supports his statements. (Beale, *Mic. in Med.*, 4th ed., pp. 232 and 259.) If the hæmoglobin is not deposited in the white corpuscles until they have reached a certain age, they will be entirely without a nucleus. If, as claimed by Böttcher and verified at this laboratory (see Quar. Mic. Journal, October, 1878, p. 46), the red corpuscle has a nucleus, the hæmoglobin must have been deposited prior to the time just given.

This time may be associated with the period when the white cell ceases its active amœboid movements, becoming passive, a condition which would occur most naturally when it was old and its nucleus small.   The appearance obtained from following the methods of Böttcher is said to be due to the coagulating effect of the corrosive sublimate on the albumen of the red corpuscle.   If this be true, it seems strange that the coagulating agent does not serve all red corpuscles alike and give a nucleus to each one.

This bleaching, hardening and staining method of Böttcher proves the existence of three classes of red corpuscles.

The red corpuscles (very few in number) having a nucleus and nucleolus, are recently derived from young white corpuscles.   Those having a nucleus only, are either from older white corpuscles or are the older forms of the red ones possessing a nucleus and nucleolus ; while those consisting of a homogeneous mass are either directly grown from the older white corpuscles, or are the oldest forms of those composing the first or second class.

The results of Beale's investigations lead to no other conclusion, and the recent researches on the structure of the nucleus by Aurbach, Hertwig, Priestley and Klein, do not, in the least, invalidate these statements.

Although there may be a difference in the structure of the red and white corpuscles, it is only such a difference as the growth of cells renders necessary.

Some reason must be given for the change in shape from a spherical body to a biconcave disc.

Hæmoglobin possesses a great avidity for oxygen, it also retains this property when united with the white corpuscles, and under proper conditions, will combine with this gas even in excess.

Will this excess of oxygen have any effect on the shape of the corpuscle ?

.   Using a carbonic acid gas apparatus, of the kind described in the *Hand-Book for the Phys. Lab'y*, by Burdon-Sanderson, and examining the blood in a suitable chamber, the effects of the gas on the red corpuscles can be studied.

It is not to be expected that the carbonic acid will unite with the red corpuscles, but the intention is to displace the excess of oxygen so far as possible, and thus reduce the red corpuscles nearer to the condition of the white.

Experiments lead to the conclusion that one of the changes resulting from this displacement of the excess of oxygen, is to render the biconcave red corpuscles more globular. The alteration is not a complete one. The red corpuscle does not become as spherical as the white, but such a complete change might be confidently expected if all the excess of oxygen could be removed. The change in form, however, is sufficient to give rise to the belief that oxygen is the active agent in causing the biconcave shape.

In speaking of the difference in color between arterial and venous blood, Foster says (Foster's *Physiology*, 3d edition, page 354, 1880) : " There may be other changes. * * * When a corpuscle swells, its refractive power is diminished. *. * * Anything, therefore, which swells the corpuscles, tends to darken blood. * * * Carbonic acid has apparently some influence in swelling the corpuscles." And it might be added, it swells them because it displaces the excess of oxygen as described above. There is no such excess of oxygen in the white corpuscles, because they have no hæmoglobin to draw oxygen to them. Dissolve out the hæmoglobin or remove the excess of oxygen from the red corpuscles, and they will not be unlike the white in shape. Hence, all that is necessary to change a white to a red corpuscle is to disseminate hæmoglobin through the substance of the latter ; this will attract an excess of oxygen, and a change in shape will result.

If the corpuscles have such a short existence, the question naturally arises : Where and how do they die ?

The serum of fresh blood contains no dissolved hæmoglobin, so that if any red corpuscles are destroyed in the circulation, either the number must be very small, or else the hæmoglobin must be speedily transformed into some other body.

Experiments made to show that the liver is a place of destruction for the red cells have given contradictory results. However, "a careful examination of the figures leads to the conclusion that the red globules are rather destroyed than formed in the liver." (*Physiology*, Küss, 2d edition, page 124, 1875.)

An account of the histology of the spleen will throw light upon the matter under consideration.   (See histology of the spleen.)

Following the divisions of the splenic artery, it it seen to divide again and again, until finally the branches diminish to the size of capillaries.   These soon become indistinct.   Cell demarcations may still be recognized, but these also soon disappear, and there is now a minute blood current without definite walls.   " As the failing branch of a drying brook wanders at last between the pepples of its bed, slender and scanty, so is it with these finest blood-currents."

The blood enters the splenic artery and flows undisturbed through its branches to the very finest capillaries.   The walls that separate it from the soft tissue now disappear, and it has to pass through a quantity of splenic tissue, with nothing to keep it from immediate contact with that tissue.

Having no walls to confine it, it flows now this side and then the other side of the "pebbles" (lymphoid cells) of its bed.   One portion of the red elements of the blood passes through this tissue into the primordial venous capillaries, and finally reaches the general circulation through the veins.

Another portion, however, meets a mechanical death by sticking fast to the splenic tissue.

The study of blood teaches that for the colored elements movement is life and rest is death. (Frey's *Comp. of Histology*, 1876, p. 121.)

The red corpuscles, being thus brought to rest, find their grave. But the younger corpuscles do not allow the older ones to remain quiet ; for, with an amœboid motion, the white cells envelop the dead bodies of the red and greedily appropriate them to their own use. In this way, the large white corpuscles mentioned above originate.

If the spleen becomes enlarged, what will be the probable result ? The larger it becomes, the more tissue there will be through which the red elements must pass, the more fine blood-currents without walls, therefore the greater the destruction of the red corpuscles. On the other hand, the number of the white cells will be correspondingly increased ; for the spleen must be considered as a birth-place of the white corpuscles. (Foster's *Physiology*, 3d ed., p. 38.)

The equilibrium will thus be destroyed and there will follow a great destruction of the red and a great increase in the number of the white corpuscles ; the extent of which will depend upon the size of the spleen.

Extirpation of the spleen does not always cause the result anticipated, and it is asserted that the number of white corpuscles is not materially changed, neither does hypertrophy of the lymphatics always follow. In answer to this it may be said that extirpation of one kidney does not always lead to any material change in the amount of urine, neither does a microscopical examination of the remaining kidney, after a time, show any increase in size of either the tubuli or glomeruli. (Flint's *Physiology*, Vol. III., 1876, p. 404.)

The spleen is classed with the adenoid tissues. (Frey, Küss, &c.) Extirpate the spleen, and, as in the case of the kidney, the remaining adenoid tissues will carry on the work.

When the spleen is removed an abnormal condition is induced and it would be difficult to assert where the red corpuscle meets its death.

Therefore, the origin of the very first red corpuscle is from nucleated cells in the vascular area ; a little later in embryonic life, from cell division.   Their origin in the adult is from the leucocytes ; the latter, becoming impregnated with hæmoglobin, owing to the action of oxygen change to biconcave discs ; the nucleus of the white cells becoming gradually changed into the formed material of the red.   Their death is owing to a mechanical cause in the spleen, and probably occurs, to some extent, in the liver also.

THEIR SHAPE.

In shape they are circular, flattened, biconcave discs with rounded borders.   When seen on the side the centre appears either light or dark, depending on the focus.   Acting as a biconcave lens, when the objective is slightly within the focus, the centre appears light, when without the focus, dark.   This led the older observers to regard this centre as a nucleus. They were " optically deluded." (See Fig. 21.) Their shape is

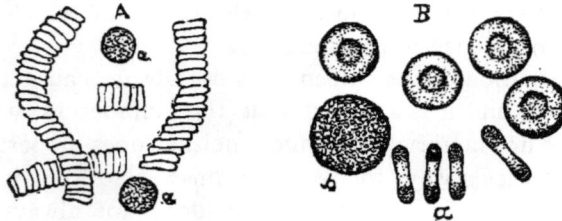

FIG. 20. A. Human blood in rouleaux. a, white corpuscles. x 400.
B. Human red blood-corpuscles. a, seen on edge. b, white corpuscle. x 1000.

readily altered by the aid of reagents and is spontaneousy changed by their removal from the circulating fluid unless especial precautions be taken for their preservation.   The red corpuscles of all the mammalia are of this shape with one exception, the camelidæ, in which they are oval.

THEIR NUMBER, VOLUME AND SUPERFICIES.

They exist in great numbers. It has been estimated that in man there are five millions of them in a cubic millimetre of blood. Welcker estimates the mean volume of a red corpuscle to be .000,000,072,214 of a cubic-millimetre, and the superficies .000,128 of a square millimetre. Taking the number of corpuscles in a cu. mm. at five million, we should have in one cubic millimetre of blood 640 square millimetres of surface. In an ordinary man of 140 pounds weight, we will say there are 12 pounds of blood, a low estimate according to many authorities. If the amount of superficies of the red corpuscles be computed for the whole 12 pounds of blood, it will give us in round numbers 38,000 square feet. Although the circulation is complete in a less time, yet a quantity of blood, equal to the whole amount in the body, passes through the lungs in not far from forty-five seconds of time. In forty-five seconds the heart will beat about fifty times. Then one beat of the heart must send into the lungs 760 square feet of surface to be oxidized. In the normal condition then there is this amount of surface of red corpuscles in the lungs at any one time, exposed to atmospheric action. We are better prepared now to understand why blood is capable of absorbing 13 times as much oxygen as the same amount of water. The red corpuscles contain less water than the serum. In 100 parts of wet corpuscles there are of water 56.5 parts, and of solids 43.5 parts. Hæmoglobin constitutes over 90 per cent. of the dried organic matter of the human red corpuscles.

THEIR STRUCTURE.

The structure of these bodies is of great interest. They are so susceptible to the action of reagents, and are so liable to undergo various changes when removed from the circulating fluid that their study is most difficult. Two questions present themselves: Has the mammalian red corpuscle an investing membrane? Has it a nucleus? In reply to the first we say : the effect of mechanical agents, the fact that at no time is anything seen at all resembling a torn or empty membrane, and the effect of heat ; the study of these makes one believe that this body is without a membrane. Believing the red corpuscle of newt's blood to possess an envelope, and knowing that the mammalian corpuscle acts toward reagents like it, Rutherford, from analogy, infers the existence of a membrane in the latter. Hensen and others are of like opinion. The majority of histologists, however, fail to find satisfactory evidence of the existence of such a membrane. It appears that the outer part of the corpuscle is more dense than the inner. It conforms more nearly to our ideas of the "formed " part of a cell.

Has the red corpuscle a nucleus ? The great majority of histologists are ready to answer positively, No. But in the *Arch. für Mic. Anat.*, Bd. 4, Professor Böttcher gives the results of some researches on this subject, confirming him in the belief first advocated by Rollet.

He uses a saturated solution of corrosive sublimate in 96 per cent. alcohol, and into fifty volumes of this solution, one of blood is to be rapidly diffused.

By this means the coloring matter of the corpuscle is taken out—bleached—and thus the internal structure brought more clearly to view. This solution preserves the corpuscles as well.

By agitating the mixture now and then the process is hastened, and in about twenty-four hours the corpuscles are allowed to subside, the superincumbent fluid poured off and pure alcohol added in like amount.

In another twenty-four hours this is poured off and distilled water added. The corpuscles are now thoroughly washed, and are not acted upon by the water.

Professor Böttcher employs eosin, hæmatoxylin, picric acid and carmine as staining agents, but prefers the first.

He finds three classes of corpuscles :

First. Homogeneous and shiny throughout.

Second. Added to this a granular mass in the center which stains readily.

Third. Besides the cortical layer and protoplasm, inclosed in the latter is a marked nucleus and a nucleolus.

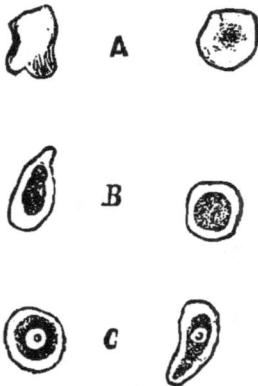

FIG. 21. The red blood corpuscles of man, treated by Bœttcher's method. A, homogeneous throughout. B, showing nucleus. C, showing nucleus and nucleolus. (Selected from Bœttcher's article, Oct., 1877, Quarterly Journal of Mic'l Science.)

Some blood was examined from a man accidentally poisoned with an alcoholic solution of corrosive sublimate with the result of finding nuclei in the corpuscles.

As soon as the public announcement of Professor Böttcher's discovery came to our notice, his method was carefully followed at this laboratory, the blood from several mammalia being examined. In nearly every case the first and second classes of corpuscles were found. It was quite rare to find those cells showing a nucleolus also. Only a few were found in all our work that appeared to be changed by the action of the reagent used. This nucleus has been demonstrated to students in the laboratory and to different members of our faculty. Various methods have been successfully used by different workers to show this nucleus. In the hands of others these same methods fail to produce the desired results. The real question at issue is this : Is the body found inside the mammalian red corpuscle, both with and without reagents, a nucleus? We are told that there is a structure characteristic of nucleus matter, an intra-nuclear fibrillar network ; but one is unable many times to see this network in that part of the cell admitted by all to be the nucleus ; and as a fact all nuclei do not seem to be possessed with it. Without any such distinctive structure as a guide how can we decide whether this is a nucleus or not? No other way appears open to us but a course of reasoning such as is advanced on page 44 on the origin of this body. The question must be regarded as unsettled, for while our own believe makes us declare in favor of the presence of a nucleus we would remind the student that such is not the generally accepted view, which is, that the red corpuscle is a homogeneous mass without membrane or nucleus.

THEIR SIZE AND MEDICO-LEGAL VALUE.

The size of these corpuscles varies not only in the same individual at different times, but also in the same drop of blood examined at any one time. The size usually given is from the $\frac{1}{3333}$ to $\frac{1}{3500}$ of an inch. The following is a list of measurements of the red corpuscles of different animals as given by Gulliver :

Dog,	$\frac{1}{3542}$	Horse,	$\frac{1}{4600}$
Cat,	$\frac{1}{4404}$	Goat,	$\frac{1}{6366}$
Hog,	$\frac{1}{4230}$	Sheep,	$\frac{1}{6355}$
Ox,	$\frac{1}{4267}$	Red squirrel,	$\frac{1}{4000}$
Brown rat,	$\frac{1}{3911}$	Black squirrel,	$\frac{1}{3841}$
Mouse,	$\frac{1}{3814}$	Gray squirrel,	$\frac{1}{4000}$

(For full table see Sydenham edition of Hewson's works, p, 237.)

What is the value of these corpuscles in criminal cases?

That is, by a microscopical examination of a blood stain or clot, either fresh or otherwise, can we distinguish human blood from that of the inferior animals? This must be considered a very easy matter in some cases. Take, for instance, the red corpuscles of all the birds, reptiles, amphibia and fishes, here they are large oval bodies with a large round or oval nucleus. The only known exception to this is in the case of the family of lampreys. In these fishes the corpuscles are circular, yet they have a nucleus very easily seen without the use of reagents, hence are readily distinguished. If, then, the question arises, "Is this human blood?" and upon examination corpuscles are found of oval shape, the answer, No, can positively be given. If, however, the corpuscles are found circular in shape, and no visible nucleus without reagents,

FIG. 22. A. a, red corpuscles viewed within the focus; b, the same without the focus. x 750. B, Frog's blood. x 400.

then an entirely different problem is involved, for with one exception (cameleidœ) the corpuscles of the blood of all the mammalia are of this shape. Here the question must be decided by measurement. First of all there must be fixed, if possible, a standard size for the human red corpuscle, for the size of the corpuscles of many of the inferior animals is so nearly like that of man that our figures in each case must give a fixed average size. Has the red corpuscle of man a fixed size? Most assuredly No, for it will be a difficult matter even to get an average size. In examining a drop of blood with high powers, one very frequently finds a few minute colored corpuscles below the $\frac{1}{4800}$ of an inch. No account will be taken of these in arriving at the size of the red corpuscle, they are easily excluded, and are few in number. The following are the average measurements of the red corpuscles of man :

Gulliver,	$\frac{1}{3200}$	of an inch.
Flint,	$\frac{1}{3600}$	"
Dalton,	$\frac{1}{3131}$ to $\frac{1}{3050}$	"
Richardson,	$\frac{1}{3318}$	"

Woodward, $\frac{1}{3082}$ of an inch.

Frey, $\frac{1}{2840}$ to $\frac{1}{4030}$ "

Welcker, $\frac{1}{3230}$ "

Our own observations give $\frac{1}{3307}$ as the mean. Thus it cannot be said that there is any settled average size, each investigator having an average of his own. While some are found as small as the $\frac{1}{5555}$ of an inch, others are as large as the $\frac{1}{2700}$ of an inch. Schmidt says, however, that over 90 per cent. of the corpuscles found in a single specimen are of the same dimensions. This much can be said, that after measuring a large number of corpuscles, if their average diameter is either $\frac{1}{3200}$ or $\frac{1}{3500}$ of an inch, or any fractional part between these two ($\frac{1}{3500}$ and $\frac{1}{3200}$) the blood may be that of man. While a number of the lower animals have blood corpuscles within these limits, (monkeys, baboons, etc., beaver, guinea-pig, porcupine, etc.), yet only the blood from certain of the inferior animals will be liable to enter a medico-legal contest.

Can dog's blood be told from human? Mr. Woodward must have settled this point conclusively. "The average of all the measurements of human blood I have made, is rather larger than the average of all the measurements of dog's blood. But it is also true that it is not rare to find specimens of dog's blood in which the corpuscles range so large that their average size is larger than that of many samples of human blood." The mean average of corpuscles in 22 drops of human blood (1766 corpuscles) ranged from .000,-309 to .000,343 of an English inch. Nearly the same number of corpuscles of dog's blood gave .000,296 to .000,340 of an inch. (Monthly Mic'l Jour., 1876, p. 132.)

Can the blood of the cat, hog, horse, sheep and ox be told from human?

Although the corpuscles of the blood of the dog and of man are so nearly identical that even in freshly prepared specimens they cannot be distinguished positively from each other ; yet the corpuscles of the blood of the animals just mentioned,—cat, hog, etc.,—are so much smaller than human blood-corpuscles that a positive distinction is possible, not only in freshly prepared specimens, but also when they are found in stains, clots, etc.

If, then, the question is asked,—Is this the blood of man as distinguished from the blood of all other animals? We shall be forced to reply, it is impossible to tell. If, however, the question is to decide between the blood of man and one of the inferior animals many times a most positive answer can be given. Between the blood of man and his most constant companion, the dog, there seems to be no difference, while it is possible to tell the difference between the blood of man and the blood of the sheep, hog, horse and ox.

To examine the stain, some of the dried blood is scraped from the surface to which it is attached, and the dried clot placed on a slide, over this is placed a thin glass cover, and a drop of the .75 per cent. salt solution being placed at its edge runs under the cover and moistens the specimen. The specimen is then examined with the highest power at command. If particles of clot are so deeply colored that the corpuscles are indistinct, their coloring matter may be washed out by a current of the salt solution. This current is easily established by placing a piece of blotting paper just opposite where the solution is applied. If now the corpuscles are too pale they can be colored. Those corpuscles most perfect in shape should be chosen, a large number of them accurately measured and their average diameter ascertained.

BLOOD CRYSTALS.

Blood crystals may readily be obtained from the rat as follows : a drop of blood is mixed with twice its volume of water and then allowed to evaporate slowly. Prismatic crystals of hæmoglobin will be seen. The blood of the guinea-pig crystallizes very easily giving the beautiful tetrahedral crystals seen in fig. 24. In the squirrel the crystals are hexagonal tables. The blood of most of the mammalia including man, yields generally prismatic or rhomboidal crystals. To obtain hæmin crystals, a drop of blood is placed in a watch crystal and about 20 times its bulk of glacial acetic acid added. The mixture is then warmed and as it evaporates the desired crystals will be formed; or to a drop of dried human blood add a few crystals of common salt, cover with a thin film of glass and place a drop of glacial acetic acid to its edge and allow it to run under and come in contact with the blood. . The specimen is then carefully warmed and soon the reddish-brown hæmin crystals appear.

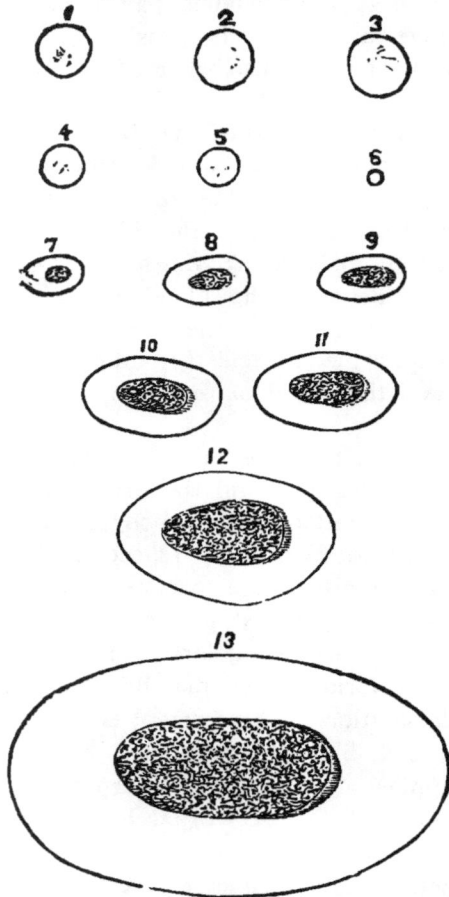

FIG. 23. Showing relative size of red blood-corpuscles of different animals. 1, man. 2, whale. 3, elephant. 4, mouse. 5, horse. 6, musk deer. 7, humming bird. 8, pheasant. 9, pigeon. 10, snake. 11, crocodile, 12, triton. 13, proteus. (Altered from Gulliver.)

METHODS OF EXAMINING.

If a drop of blood be placed on the slide, covered at once with the thin glass and transferred to the microscope, the examination will be anything but satisfactory. The number of corpuscles in the field is so great and the number of layers so many that the specimen cannot be studied to advantage. Again, if mixtures are used to dilute the blood, unless prepared with the greatest care, they will cause changes in the corpuscles. When but a comparatively short examination is required, the following method has always been highly satisfactory to us and is constantly employed in this laboratory. To procure the drop of blood, one of the fingers is congested by tying around its base a string or handkerchief. When well filled with blood a fine cambric needle is quickly thrust through the outer coats of the skin over the end of the congested organ. One surface of a glass slide is now gently breathed upon and a drop of blood, pressed from the puncture, is brought in contact with this slightly moistened surface.

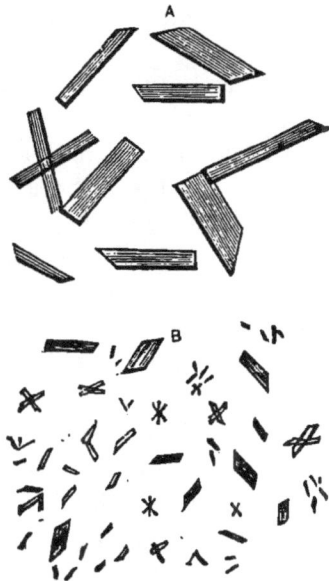

FIG. 24. A. Hæmatin crystals, human.
B. Hæmin crystals, human.

One surface of the cover glass is now breathed upon and immediately its edge is placed close to, just in contact with, the edge of the drop of blood. With the aid of a needle the cover is lowered away from the drop,—not over it,—until it touches the slide. The blood corpuscles will readily flow between these moist glasses, by capillary attraction, until the surface beneath the cover glass is nearly or entirely covered.

There is but one single layer of corpuscles and they show to the best possible advantage. For this method to be successful it should be carried out rapidly and the moisture should not be

in excess, lest the water cause some change ; it should be suf-
fficient, however, to allow the corpuscles to flow readily under the
cover.   The amount of moisture to be impartedby the breath is
soon learned after a trial or two. If desired, a layer of oil may be
placed around the cover glass and thus prevent the drying of
the specimen.

Examined with a magnifying power of about 450 diameters one never fails in a specimen prepared as above to see
most beautiful and perfect representations of the rouleaux,
where a greater or less number of red corpuscles adhere together by their flat sides. A few corpuscle will be seen in the
field alone, the majority of these few have a different appearance from those in rouleaux. They are larger, globular, and
more granular. They are the white blood corpuscles or leucocytes. If the cover-glass be touched lightly with a needle
the red corpuscles will roll about readily and their rouleax will
be destroyed, while the white ones do not roll about readily
and appear as islands in the midst of a rapid stream. Another
satisfactory method of obtaining a thin layer of blood is to
draw the edge of a smooth glass slide over the fresh blood and
then drawing this edge of the slide across the flat surface of
another slide. The thin cover is at once applied and the specimen examined.   Reagents can be applied by simply touching
a small drop to the edge of the cover glass.

ACTION OF REAGENTS.

Water has the effect of changing the red globules to a
spherical form. Many times the central depression of one
side will disappear before it does on the other, giving them
a peculiar cup-shaped form; finally the water dissolves out
the coloring matter, the corpuscles become paler and paler
until after a time they are totally disintegrated. Acetic acid
brings about these changes with great rapidity. The alkalies
are very active in dissolving their whole substance.   A 1 p. c.
solution of salt will cause the red corpuscles to assume the
crenate or horse-chestnut form, in which their surface is

covered with minute spinous projections. This same appearance is seen in corpuscles that have been exposed to the air a short time. This may be due to the loss of carbonic acid by the plasma, the corpuscles then loose carbonic acid themselves and this is followed by a shrinking of part of the stroma, probably the net-work of fibrils. (Klein.) Electric discharges of the leyden jar cause them to assume the crenate form, to finally swell up and become deprived of their color. Tannic acid, in a 2 p. c. solution causes the hæmoglobin to collect at the periphery in the form of one or more round masses.

FIG. 25. A, red blood corpuscles after the action of water. B, the same after the action of salt solution.

THE WHITE CORPUSCLE.

Besides the red corpuscles there are seen other cellular bodies which are unlike the former in shape, size, structure and color.

As seen under the microscope they are destitute of any color, hence they are called colorless corpuscles, or white corpuscles.

In nearly every specimen of blood some white corpuscles will be found smaller than the red, but the great majority are larger, not only in man, but in all the mammalia, while in other animals,

FIG. 26. Pus. A, before, and B, after, the action of dilute acetic acid.

owing to the large size of the colored corpuscles, the white are the smaller. They are about the same size in all the verte-

brates, and although varying in the same and different specimens, yet their diameter will not vary far from the $\frac{1}{2300}$ of an
inch. When dead, or in a quiescent state they are nearly
spherical, but when living they are capable of altering their
shape in a most remarkable manner, their substance projecting
into prolongations which in turn are retracted into the body of
the cell after the fashion of our amœba (page 30.) The white
corpuscles of the blood of the newt can be seen to undergo
these amœboid movements without any extra care on the
part of the operator.   In the case of those from human blood
it will be necessary to have the specimen maintained at, as
nearly as possible, the same temperature as that of the body,
viz., 98.5 degrees F.

The number of white corpuscles does not bear any constant relation to the number of the red.   Before meals the
number of white to red is about as 1 to 300.   In a fasting condition the number may decrease until the ratio is as 1 to
1000.   In certain pathological conditions the number of white
corpuscles is increased until they nearly, if not quite, equal
the red.   If fluids be taken freely with a meal the number of
white corpuscles is greatly increased, while if a meal be taken
without drink the number may remain the same or even be
diminished.

If a current be established through the specimen the red
corpuscles can all be washed away and the white left alone in
the field.   They appear as highly transparent bodies, dotted
with minute granules.   In each globule there is one nucleus,
capable of amœboid movements independent of the body of
the cell.   Often two or three nuclei with nucleoli are present.
Heitzman makes the white corpuscle composed of a minute
net-work of fibrils,—the intra-cellular net-work—in the meshes
of which there is an interstitial hyaline substance ; and the
granular appearance is simply these minute fibrils seen in
optical section.   The nucleus also is composed of an intra-

nuclear fibrillar net-work which is directly connected with the intra-cellular. This appearance can be seen only under the most favorable circumstances.

The large granular corpuscles seen in human blood consist of a deposit of real granules in the meshes of this intra-nuclear net-work. These granules may be the broken-down remains of what once constituted the minute fibrils of the corpuscles.

The white corpuscles are continually being removed or changed into other forms, for their number varies extremely under different circumstances and at different times. Whence the supply?

We find the lymph constantly pouring into the blood vast numbers of them which appear for the most part after the lymph has traversed the lymphatic glands. In a section of a lymphatic gland are seen corpuscles identical with the white corpuscles of the blood. They are of different sizes and have dividing nuclei. Klein says that he has seen these corpuscles budding off from the reticulum of the spleen. There origin then is chiefly from the lymph corpuscles of the lymphatic glands. Other sources may exist, as from the endothelial cells of serous membranes, and from almost any proliferating tissue.

ACTION OF REAGENTS.

Dilute acetic acid increases their transparency and renders the nuclei more perceptible. Its final effect is the same as that produced by water, viz.: their total disintegration.

Alkaline solutions if strong enough dissolve them.

A slight electric shock causes the living corpuscle to contract, while a strong shock has the same effect as the three reagents just given.

CHAPTER IV.

Epithelium and Hair.

UNDER this head we include the crowded cells covering mucous membranes and the skin, also the cells lining secreting and allied glands. Epithelium may be divided into two classes according to the form of the cells. 1. Columnar, in which the cells are long and narrow. 2. Pavement, in which the cells are flattened. As varieties of the first class we have cylindrical, ciliated, etc., and of the second class scaly, buccal, flattened, etc.

Epithelium may be arranged in a single layer as on the villi of the intestine or in stratified layers as on the surface of the body. When stratified layers of epithelium occur it is almost invariably of the pavement kind, although this class appears as a single layer in many parts of the body. Cylindrical epithelium rarely presents more than a single layer.

If epithelial cells be separated from their natural connections by the aid of the clean blade of a knife and placed with an indifferent fluid on a glass slide, covered, and examined with a high power, cells will be found isolated and favorable for study. They appear to be composed of a soft albuminous matter in which is a large number of "granules." Under most favorable conditions one is able to see what Hutzmann, Eimer, and Klien first taught, namely, that the body of the cell is composed of a minute intra-cellular net-work of fibrils, in the meshes of which is a hyaline substance. In the ordinary pavement variety this fibrillar net-work is irregularly arranged, while in the columnar variety the intra-cellular fibrils have a longitudinal arrangement, parallel to the long axis of the cell.

According to this view, the familiar dots or small granules are these intra-cellular fibrils seen in optical section. Some of these cells certainly appear to have a membrane; this may be the hardened part of the cell from evaporation, or more probably the dense exterior of its formed material.

With the exception of those cells on the surface of the body, each cell has a nucleus surrounded by a membrane, and composed of an intra-nuclear fibrillar net-work. In its meshes we find an interstitial substance. The granules are due here to the same cause as those found in the body of the cell; some-times however they represent the broken-down remains of fibrils when they are true granules. Inside the nucleus is seen frequently a nucleolus, from which the nucleus is developed. A refractile body, seen in the nucleus at times and resembling a nucleolus, is found, upon closer examination, to be a part of the shrunken intra-nuclear net-work. The cells are held in close contact with each other by an albuminous substance termed "cement substance." In examining transverse sections of hardened specimens, this substance looks like a thin mem-brane separating the cells, but in the fresh specimens it is clear and viscid. Many of these cells are being constantly discon-nected from their bases by the acts of nature. The old cells are ever falling off spontaneously and as the result of the pressure and friction to which many surfaces of the body are subject. Some of them seem to be more enduring in their nature as those found in different parts of the eye.

CILIATED EPITHELIUM.

Certain of the columnar cells have fine hairs on their bases, called cilia. These cilia are but the prolongations of the fine protoplasmic filaments,—the intra-cellular fibrils—through the base of the cell. The regular vibrations of these cilia led the old observers to discuss electrical attraction and repulsion. Klein explains this motion as follows:

He supposes that the "intra-cellular net-work contracts to one side in a horizontal diameter; each such contraction acts naturally on the lower ends of the cilia, which are pulled thereby to the same side, while the outer or freely projecting portions of the cilia are driven in the opposite direction. Each cilium represents a lever, the short arm of which is within the cell in connection with the intra-cellular net-work, the long arm being the freely projecting part and the fulcrum or fixed point lying in the membrane covering the free cell border. The next moment the contraction of the intra-cellular net-work ceases and the cilia move again in the opposite direction."

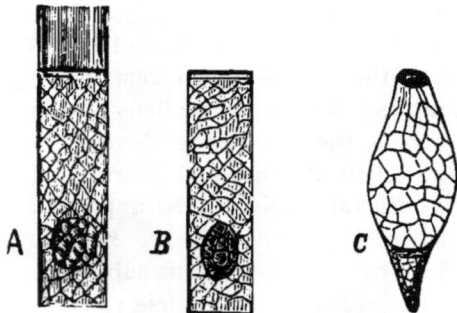

FIG. 27, A, Ciliated cell. B, Columnar cell. C, The columnar cell, B, changed into a goblet cell and filled with mucin. (after Klein and Smith.)

METHODS OF EXAMINING.

For samples of columnar epithelium remove a portion of the intestine of a recently killed rabbit or cat, and wash carefully the mucous membrane by flowing over it some of the salt solution. Then with the point of a scalpel transfer a small portion of the membrane to a glass slide and to it add a drop of staining fluid. Certain of the villi will be removed entire, but a few of the columnar cells will be seen floating free in the field. At fig. 28 are seen a few of these cells. An oval nucleus is present in each cell bounded by a distinct line. These cells terminate usually in a fine point, a few are bifid, while others terminate in a rounded extremity. At the free border of the cell is a thickened margin in which fine vertical striæ are seen.

If the animal be killed during digestion, globules of fat within the cells will be recognized by their strong refractive power. A few cells may be seen of peculiar chalice shape called "goblet cells." The part of the cell nearest its free border has swollen out owing to the conversion of its interstitial substance into mucin. The membrane at the base will finally rupture, and the mucous contents be poured out. The nucleus will be found toward the distal end of the cell. Hæmatoxylin stains the mucous contents of these cells a deep purple-blue. Pieces of the small intestine of the cat or rabbit may be hardened in a one per cent. solution of potassic bichromate or osmic acid. In a few days a portion of the mucous membrane may be removed and examined in the salt solution. All parts of the cell now show to better advantage. The fatty particles in the cells are colored black by the osmic acid, and the specimens can be permanently preserved in glycerine. If the free surfaces of the cells be turned toward us, they present a beautiful mosaic, caused by their collateral pressure during their growth.

FIG. 28. Epithelium from intestine, a, goblet cells. x 400.

Impregnation with silver is to be especially recommended in studying the epithelia. For this purpose the specimen is immersed in the solution for a short time, perhaps the fractional part of a minute, when it is removed, washed in water and exposed to the light until it assumes a brownish color. The specimen is then examined in acidulated water or in dilute glycerine. This colors the cement substance, the cell itself not being affected, and the boundary lines become very distinct. The cells may be afterwards stained with carmine if desired. Very weak solutions of the silver (argentic nitrate) are employed, from .2 to .5 per cent. and in some cases even weaker solutions can be more advantageously used.

CILIATED EPITHELIUM.

This may be readily obtained by gently scraping the mucous membrane on the roof of a frog's mouth. If a large amount be desired, the frog may be killed and this entire membrane dissected off. For ordinary purposes sufficient epithelium will be brought away with the mucus by scraping the membrane *in situ* with the point of a clean scalpel. A drop of normal saline solution has been placed on the centre of a glass slide and on either side of it a hair, in order that the cover glass may not press upon the cilia and thus check their vibrations. The adherent mucus is now transferred from the knife to the salt solution. The preparation should be covered and examined at once with a power of at least 450 diameters. The cilia will be seen in active motion, driving any particles that come near them in one direction. If the cells are isolated or if only a few are in a group, the cilia will act like little paddles, causing the cells to whirl around and move about over the field. If there are no blood-corpuscles or granules to be driven about, particles of carmine may be added to the specimen. This is best done by mixing a little granule of carmine with water and placing a drop to the edge of the cover glass. Diluted with water aniline red will tinge the cilia without causing their motion to cease. (Aniline red=Fuschin, 1 centigramme. Absolute Alcohol, 20—25 drops. Water, 15 cubic centimetres.) The action of cilia may be strikingly shown by examining the gills of a living oyster or of a clam.

This variety of epithelium is found in the following places in the human body: Throughout the upper part of the nasal passages, nasal duct, posterior surface of soft palate, upper part of pharynx, larynx, trachea, bronchial tubes to near their termination, eustachian tube, tympanic cavity, internal surface of eyelids, ventricles of brain, central canal of spinal cord, fallopian tube, body and, according to some authors, neck of uterus, vascula efferentia, coni vasculosi, and canal of epididymis, and upper half of vas deferens.

ACTION OF REAGENTS.

Electrical currents, heat, and any fluid currents all accelerate the motion of the cilia. Carbonic acid first accelerates, then checks, and finally arrests their action. When the motion has just ceased slightly alkaline fluids will apparently call it to life again for a short time.

To obtain squamous, or buccal epithelium move the tongue roughly over the gums and interior of the mouth, then place a large drop of the saliva on the slide. The air bubbles may be removed by passing a needle horizontally over the specimen skimming them off ; cover and examine.

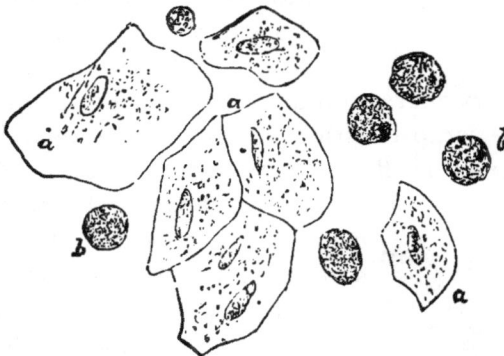

FIG. 29. Ciliated epithelium from uterus (human). x 400.

FIG. 30. Saliva. a, epithelial cells. b, salivary corpuscles. x 400.

The cells are partly isolated, partly hanging together. Their large size, 1-450 to 1-750 of an inch, and their single small

nucleus are characteristic.   A varying number of salivary corpuscles is seen in every specimen.   They are rather larger
than the white blood-corpuscle and there is a peculiar Brownian
movement of their enclosed granules.   One is inclined to believe, however, that these are exuded white blood-corpuscles
swollen by the watery constituents of the saliva.   In examining the saliva one frequently finds remains of food, as fibres of
meat, starch granules, etc.

Pigmented pavement epithelium may be found covering
the choroid, ciliary processes and posterior surfaces of the
iris.   Shreds of it may be removed with a scalpel and examined
as usual.   Here again the cells form a beautiful mosaic of a
hexagonal form.

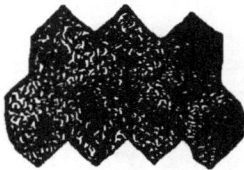

FIG. 31.   Pigment epithelium.
(Choroid.)   x 400.

The pigment granules may be so
numerous as to entirely obscure the
nucleus, while in other cells only a few
of the melanotic molecules are embedded in the soft substance.   The glandular and allied varieties of epithelium
will be studied in connection with the
different organs. The surface of the body
is covered everywhere with a stratified layer of epithelial cells.
If the skin be scraped with a scalpel a fine dust is obtained,
which, examined in fluid, is found to consist of irregular,
broken epithelial cells without a nu-
cleus.   These layers of cells are best
studied in a section of skin.   (See organs of sense.)   Closely allied to these
cells are those found in human nails,
which commence the third month of
fœtal life as an elevation of the skin of
the distal phalanx.   The nail substance
is composed of epidermal cells very
closely united, but easily separated by a

FIG. 32.   Epithelium from the
back of the hand.   a, from surface.   b, deeper   x 400.

number of chemical reagents. These cells are irregular in shape and enclose a round or lens-shaped nucleus. A 27 p. c. solution of potash is one of the best reagents to isolate the

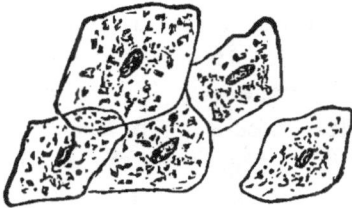

FIG. 33. Epithelium from the nail   x 400.

cells and to exhibit their nuclei. By boiling thin sections of nail in a 10 p. c. solution of soda the individual cells are demonstrated very quickly. Berthold proved the life of a nail cell to be four months in the summer and five in the winter.

HAIR.

Close in this connection comes the hair and its tissues. Hairs cover nearly the entire surface of the body, varying in size and physical characters in different situations. The hair is ordinarily coarser in women than in men, dark hair coarser

FIG. 34. Human hair (white.) x 500.

than light. Hairs are very elastic and may be stretched one-third more than their entire length. They are also very strong, one from the head will on an average bear a weight of six or seven ounces. They can become strongly charged with negative electricity by friction. They readily absorb moisture becoming sensibly elongated. In a human hair cut from the head three parts are distinguished. First, the cuticle, which is composed of thin, flattened, quadrangular cells arranged in an imbricated manner. The outlines of these cells give to hair its peculiar markings. Second, the cortex, or fibrous substance. This is best seen in a hair that has been boiled in sulphuric acid and then brushed.   A few longitudinal lines resemble fibres ; they are long, irregular, or spindle-shaped, flattened cells.   Third, the medulla.   This is present only in a portion of the hairs. It is not found in the soft, downy hairs of the body, and is not always present in the hairs of the head.   In size it is equal to one-fourth or one-third the diameter of the hair. It is composed of small, round or polyhedral nucleated cells.   In these cells are sometimes seen granules of pigment of various colors, although the mass of coloring matter appears to be in the cortex.   In the medulla are small air globules between the innumerable small cells.   The color of the hair depends upon the color and amount of pigmentary matter in the cortex and the amount of air in the medulla.   In hair that has become gray from old age there is a marked loss of pigment, while in sudden blanching of the hair the amount of pigment remains normal but the medulla becomes filled with air, some entering the cortex also.

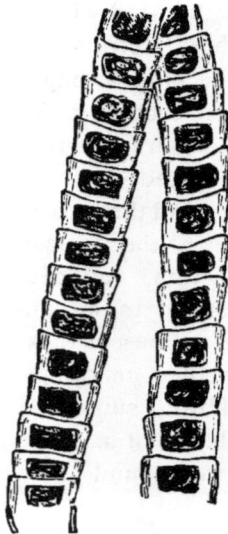

FIG. 35. Cat's Hair.   x 400.

The hair as it lies in its oblique sac in the skin has surrounding it an involution of the corium, and frequently that of the subcutaneous cellular tissue.   The outer part of the sac is composed of longitudinally and transversely arranged connective tissue.

Internally is the hyaline boundary layer. At the bottom this layer forms a papilla from which the hair is formed and nourished.   The hair bulb rests on the papilla. As the skin is involuted we find its horny layer dipping down in the sac to make the internal root sheath, while the rete-mucosum forms the external sheath.   The horny or internal root sheath divides into two layers, the superficial layer consists of

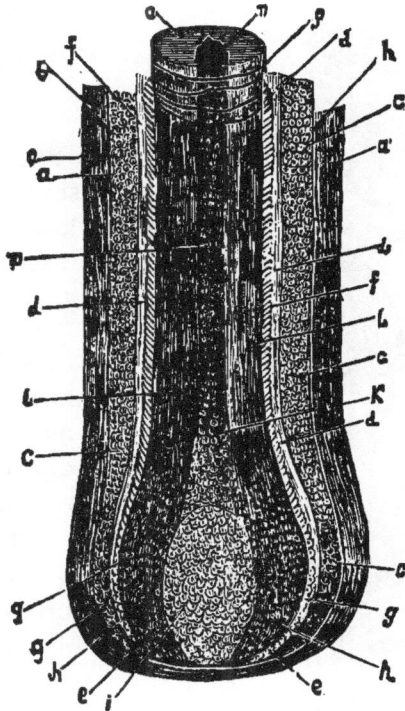

FIG. 36.—Human hair. a, the sac. b, its hyaline inner layer. c, the external, d, the inner root-sheath, e, transition of the outer sheath to the hair-bulb. f, epidermis of the hair (at f, in the form of transverse fibres.) g, lower portion of the same. h, cells of the hair-bulb. i, the hair papilla. k, cells of the medulla. l, cortical layer. m, medulla containing air. n, transverse section of the latter. o, the cortex. (Frey.)

vertically arranged non-nucleated cells, and the deep layer consists of cells which are nucleated and arranged around the hair shaft. While the hair is still within the sac there is a double layer of hyaline cells standing obliquely around it. The outer layer terminates with the sac, but the inner one covers the shaft of the hair throughout its entire length. (See above.) Sometimes the inner layer presents the appearance of transverse fibres. Hair grows by a cell increase from the lower part of the hair bulb. It commences at the

FIG. 37. Transverse section through a human hair and its follicle. a, hair. b, epidermis of the same. c, inner, and d, outer layer of the inner root-sheath. e, outer root-sheath. f, its peripheral layer of elongated cells. g, hyaline membrane of the hair sac. h, its middle, and i external layer. (Frey.)

end of the third or beginning of the fourth month of fœtal life.

To obtain transverse sections of hair it has been recommended to shave the chin and then in an hour repeat the process. This has proved a very poor method in our hands. The sections are cut obliquely, and are very unsatisfactory for study. A much better method is to tie a number of hairs together and dip them in a solution of gum arabic. Remove at once and when the gum is dry imbed the bunch and cut the sections, or a few hairs may be placed between two pieces of pith and very satisfactory sections cut off-hand with a sharp razor. For studying the hair sacs sections of the scalp are prepared.

CHAPTER V.

Connective-tissue Group.

WHITE FIBROUS — YELLOW ELASTIC — ADIPOSE TISSUE — PIG-
MENT CELLS—CARTILAGE—BONE.

WHITE FIBROUS.

WHITE fibrous tissue is extensively diffused throughout the body. It is arranged in the form of bundles which have a wavy appearance more or less marked according to the degree of contraction. These bundles may be small and thin, or very large and thick and possessed of considerable elasticity. Each bundle is composed of minute elementary fibrils which are held together by a "cement substance." Acids and alkalies destroy the fibrous appearance of this tissue by converting the fibrils into gluten or gelatine. In some parts the bundles of connective tissue are surrounded by a hyaline sheath so that when reagents are applied and cause the bundles to swell up, the sheath is torn into transverse portions which rapidly contract between the protruding portions of the bundles. This gives the peculiar rings or points of constriction.

The primitive fibrils of our bundle do not divide, neither do they anastomose. We find them arranged differently in various parts of the body, loosely as in the subcutaneous tissue, and then firmly in parallel groups as in tendon. Two kinds of cells belong to connective tissue. First the mobile, and second the fixed. The mobile, or migratory cells, are lymphoid elements which have left their proper channels in the blood and

73

lymphatics to slowly wander through the tissues.   The fixed
connective tissue cells consist of an oval nucleus surrounded
by a thin structure of protoplasm which extends out into points

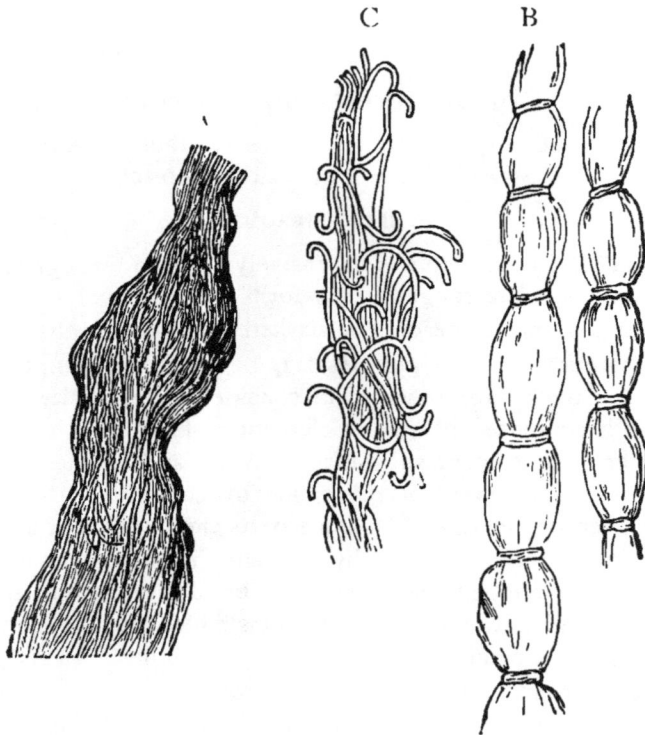

FIG 38   A, White Fibrous. B, the same after adding acetic acid.
C, Yellow Elastic. x 400.

at the periphery.   These processes may be few or many in
number, short or long, and by these the cells are brought into
continuity with each other so as to form a net-work.   Waldeyer
describes a large, coarse, nucleated, granular cell found near
vessels, especially the arteries and called by him "plasma
cells."

METHODS OF EXAMINING.

A small piece of subcutaneous tissue is placed on a slide and moistened with the salt solution. Bundles of the fibrils are seen with their wavy appearance. A few elastic fibres may be seen here and there in the field. Acetic acid will cause this fibrous appearance to disappear and the tissue to look like a mass of jelly. The cells are seen without much difficulty in a specimen of intermuscular fascia. A small portion of this is excised with the scissors and by the aid of needles carefully spread out on a dry slide; the specimen is kept moist by occasionally breathing upon it. A drop of hæmatoxylin is now placed on a cover glass which is inverted upon the specimen. A drop of acetic acid is placed at the edge of the cover and the cells now show to good advantage. Specimens of this tissue may be teased from a tendon while fresh or the tendon may be hardened in chromic acid and longitudinal and transverse sections examined.

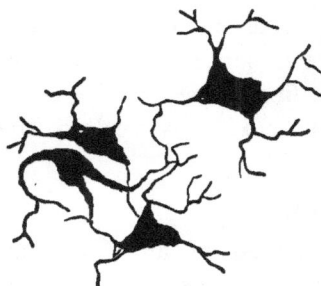

FIG. 39. Connective tissue cells from the perimysium of an ox. x 400.

White fibrous tissue is developed from embryonic connective tissue cells which are spherical at first, but afterwards become elongated, then spindle shaped with a nucleus. The protoplasm of the original cell is directly transformed into a bundle of connective tissue fibres, the nucleus gradually disappearing. (Boll.) Henley describes another method where the embryonic cell produces a peripheral substance in which subsequently bundles of fibrous tissue are formed.

YELLOW ELASTIC.

Yellow elastic tissue is composed of distinct, round, individual fibres which anastomose, divide, have a tendency to

curl at the ends, and are not affected by acetic acid.   In this
last respect they are in direct contrast to the white, this enables
them to be recognized very easily.   By their dividing and anastomosing an elastic net-work is formed.

METHODS OF EXAMINING.

These fibres are found of large size in the ligamentum
nuchæ of a calf or ox.   A small piece is separated thoroughly
with needles and examined.   These fibres are developed directly from nucleated embryonic cells.   If it is desired, they
can be preserved in glycerine.

ADIPOSE TISSUE.

Adipose tissue is generally possessed of an 'artery and
vein and a rich net-work of capillaries.   This net-work is usually small, in most cases surrounding but one, two, or three of
the fat cells.   The cells are aggregated into groups, each group
having its afferent and efferent vessels.   A fat cell is a spherical body containing a large fat globule which occupies the
bulk of the cell.   The wall of the cell is very thin and surrounds the fat on every side.   In the cell
is seen occasionally a nucleus ; this is
many times obscured from sight, and
is only recognized after the fat has been
dissolved out of the cell.

The connective-tissue corpuscles are
transformed into fat cells by having deposited into their interior small fat globules which increase in size and number
until they become confluent in one or two
large drops.   As a result of this the size
of the cell is greatly increased.   When
the adipose tissue in the body is being
reduced from starvation, or from other causes, the fat
disappears from the cells and a clear fluid takes its

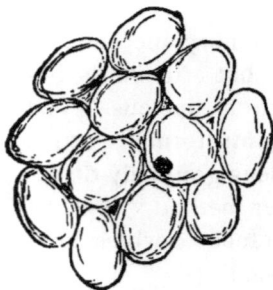

FIG. 40. Adipose Tissue.
x 200.

place. After a time this too may be removed and the cell return to its original connective-tissue corpuscle.

METHOD OF EXAMINING.

To study this tissue it is only necessary to place a small piece on the slide, cover and examine, first with low and then with higher powers.   Injected specimens are much to be preferred.   Thin sections are cut from the fresh specimen by the aid of the freezing microtome.   Sections can be mounted in glycerine ; if injected, in Canada balsam.

PIGMENT CELLS.

These are much more common in the lower than in the higher vertebratæ.   In the lower animals the connective-tissue cells are filled with pigment granules of various colors from a jet black to a greenish or gray.   Here the cells have long processes which anastomose with each other.   The processes constitute about all of the cell, only a nucleus is present where the body of the cell is usually found.   These cells are capable of altering their shape, for when subjected to certain irritants they can withdraw their pigment processes entirely, becoming changed into a round, spherical body with a central nucleus.   This change is under the control of the nervous system and accounts for the rapid changing of color observed in many animals.   It will be understood, however, that the contraction of the pigment processes does not necessitate the contraction

FIG. 41.   Pigmented connective tissue cells from mammalian eye. x 400

of the whole process of the cell. Only a portion of the process of the cell is occupied by pigment.   The part containing the pigment is the intercellular fibrils (Klein) and the part not containing it, the matrix of the cell.

In man, pigment cells are limited almost exclusively to the eye.

The cells with their long processes described above are always seen when examining the circulation of the blood in the web of a frog's foot.

CARTILAGE.

There are three varieties to be studied. First, the hyaline with a homogeneous interstitial substance. Second, the fibrous, in which the matrix is split up into fibres. Third, the reticular. The articulating surfaces of bone are covered with a layer of cartilage from $\frac{1}{50}$ to $\frac{1}{25}$ of an inch thick. In this situation it is not covered with a membrane, but with this exception all cartilage is covered with a thin vascular membrane called perichondrium.

Hyaline cartilage is composed of nucleated spherical, or oval cells in which are occasionally seen small fat globules, and of a hyaline matrix. Lacunæ are seen of an oval or round shape measuring from $\frac{1}{1200}$ to $\frac{1}{350}$ of an inch. They are lined with a membrane which has been demonstrated to possess minute openings (Arnold). These lacunæ are filled completely by the cells in living cartilage, but after death, and by the use of reagents, the cells shrink away from their walls and a space is left between them and the walls of the lacunæ. These cells may be quite close together or they may be separated some little distance by the matrix substance. In growing cartilage the nuclei of the cells divide, afterwards the whole cell, until a number of cells is produced. These cells soon become separated from each other by a considerable amount of matrix derived from their growth. The matrix of this variety is a homogenous hyaline substance which yield "chondrin." It is firm, structureless, without blood-vessels or nerves, and is derived from the cells themselves. It is permeated all through its substance with fine channels which anastomose with the openings in the walls of the lacunæ (Arnold). These spaces are the lymph channels of our tissue. In certain pathological conditions, and in old age, the salts of lime are deposited in the matrix of this cartilage. Ranvier tells us this process commences next to the cartilage cells. By the aid of certain reagents the homogeneous appearance of the ground-substance of this tissue is destroyed, and it is proved to consist of thick rings surrounding the individual cells or groups of cells, proving without doubt

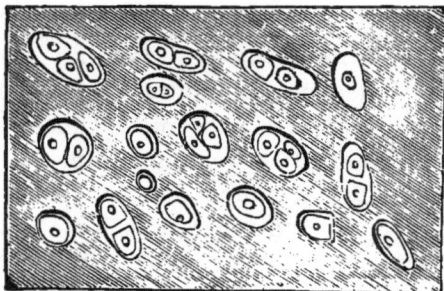

FIG. 42. Hyaline Cartilage. x 400.

that this part of the cartilage represents the formed material of the cells.

Fibrous cartilage as found between the vertebræ, symphysis pubis, etc., differs from the first variety in that its matrix is composed of bundles of ordinary fibrous tissue. At the point where tendinous tissue passes into fibrous cartilage the fibres of the former pass uninterrupted into those of the latter.

FIG. 43. Thyroid cartilage of the swine. The basis substance is divided into cell-districts by means of chlorate of potash and nitric acid. (Frey.)

In yellow elastic or reticular cartilage as found in the lobe of the ear, larynx, epiglottis, etc., the ground-work consists of hyaline cartilage which is permeated by elastic fibres. These anastomose and divide to form a reticulated framework. In this variety the cells are more abundant and surrounded by a homogeneous area.

METHODS OF EXAMINING.

For the study of hyaline cartilage nothing can excel the preparations obtained from the thin cartilage projecting from the sternum of a recently killed young newt. Any tissue covering the cartilage is easily removed and many times the specimen is thin enough for immediate examination. It is moistened with a drop of the normal saline solution and examined with a ¼ inch objective. Remove the salt solution and add a drop of a 5 p. c. solution of acetic acid. Now the nuclei become more granular and distinct.

Most beautiful specimens are prepared as follows :

A thin section is placed on a slide and covered with a drop of hæmatoxylin, which is allowed to remain two or three minutes, then it is washed off with alcohol. A drop or two of acetic acid is added and in a moment washed off with alcohol. The strength of the acid used will depend upon the degree of coloring imparted to the tissue by the staining. If the coloring was deep, the acid should be of full strength, if not so deep, then from a 10 to 20 p. c. solution should be used. By this method the matrix is stained but slightly, the cells more so, and the nuclei most intensely. The spaces between the cells and matrix are now very distinct and the whole specimen shows everything to be desired. The sections are best preserved in glycerine.

Chloride of gold is here highly recommended. The section is placed in a 1 p. c. solution for 15 or 20 minutes, then exposed to the light in distilled water for 24 or 36 hours, and finally mounted in glycerine. The cells are stained violet and they are not caused to retract from the matrix, the nuclei are colored a reddish tint and the matrix is scarcely stained at all.

Osmic acid is useful in that it stains all fatty particles black.

A .25 p. c. solution should be employed and the section allowed
to remain in it for 12 or 14 hours.

With the other varieties of cartilage the chloride of gold
method may be employed or they may be examined fresh.

To show the layers arranged concentrically around the
cells, the cartilage may be digested in water at about 100 degrees F., or dilute sulphuric and chromic acids, or a mixture
of nitric acid and chlorate of potash.

BONE.

In man bone forms the whole of the skeleton and the cementum of teeth.   We find this true in most of the vertebrata.
Histologically we distinguish two parts in this osseous tissue,
the matrix and the bone cells.   This distinction is easily perceived, if a thin section be placed under the microscope and
examined by transmitted light.   The matrix is firm and brittle

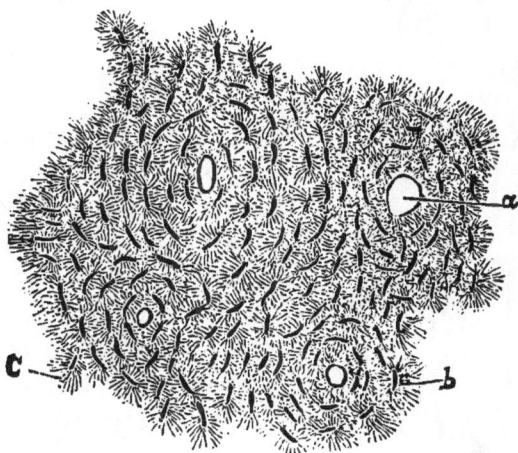

FIG. 44. Transverse section of bone (man,) a, Haversian canals. b, lacunæ.
c, canaliculi. x 50.

and impregnated with insoluble inorganic salts. By the action
of dilute acids the carbonic acid is eliminated from its combi-

nation with the lime which is rendered soluble, while the bone becomes soft without changing its form. The soft matrix represents the "ossein," which is composed of minute fibrils arranged parallel to, or interlacing each other.

The bone corpuscles are lacunæ with long branches, canaliculi, by means of which they may anastomose with neighboring lacunæ. Each lacuna contains a nucleated cell of protoplasm, the bone cell proper. For histological study bone may be divided into two classes, compact and spongy.

In compact bone we recognize ; I, Haversian canals. II, bone lamellæ. III, bone cells.

I. The Haversian canals are for the purpose of conveying blood-vessels and lymphatics. They vary in size, but average about the $\frac{1}{500}$ of an inch in diameter. They may be round or oval in shape. Where the bone is most compact near the outer surface they are very small, but towards the central cavity they acquire a large size. In a longitudinal section these canals may be seen to form elongated meshes, communicating with each other by branches given off at acute angles or more generally by means of short oblique branches. They open either upon the external compact substance of the bone or into the central

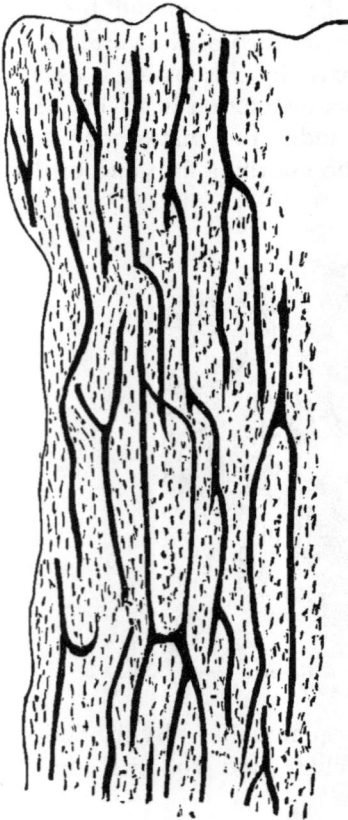

FIG. 45. Longitudinal section of bone showing bloodvessels. (Haversian canals.) x 25.

medullary spaces. They are lined with a delicate membrane. Around these Haversian canals are seen in transverse sections, layers of rings termed lamellæ.

II. These are arranged as follows : 1. Concentrically. As many as fifteen lamellæ are occasionally counted around one canal, but the number varies exceedingly, the smaller

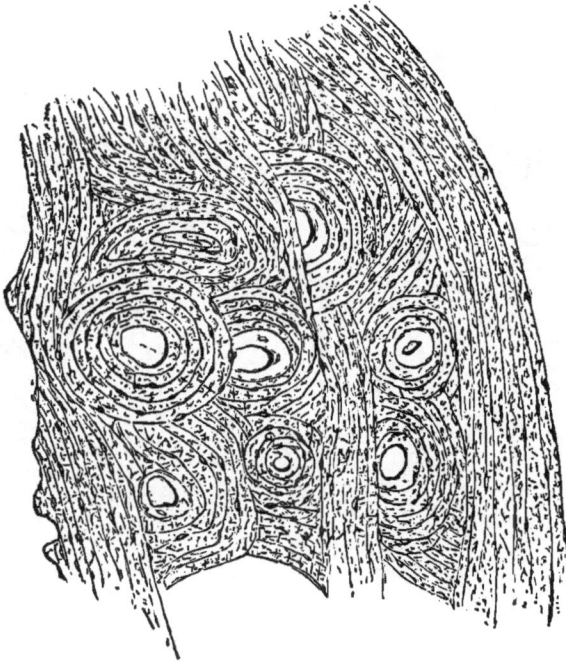

FIG. 46. Transverse section of bone, showing lamellæ. x 50.

canals having fewer than the larger ones. 2. Interstitial lamellæ. These are more or less curved and run in various directions. 3. The circumferential lamellæ are disposed parallel to 'and in contact with the periosteum externally and limiting internally the large medullary canals.

III. The bone corpuscles are found between the lamellæ and are arched to correspond with their curve. They
are quite numerous. Welcker gives on the average 740 to
the square millimetre, and Harting 910. They appear as
dark, black figures with a central body, lacuna, and branched

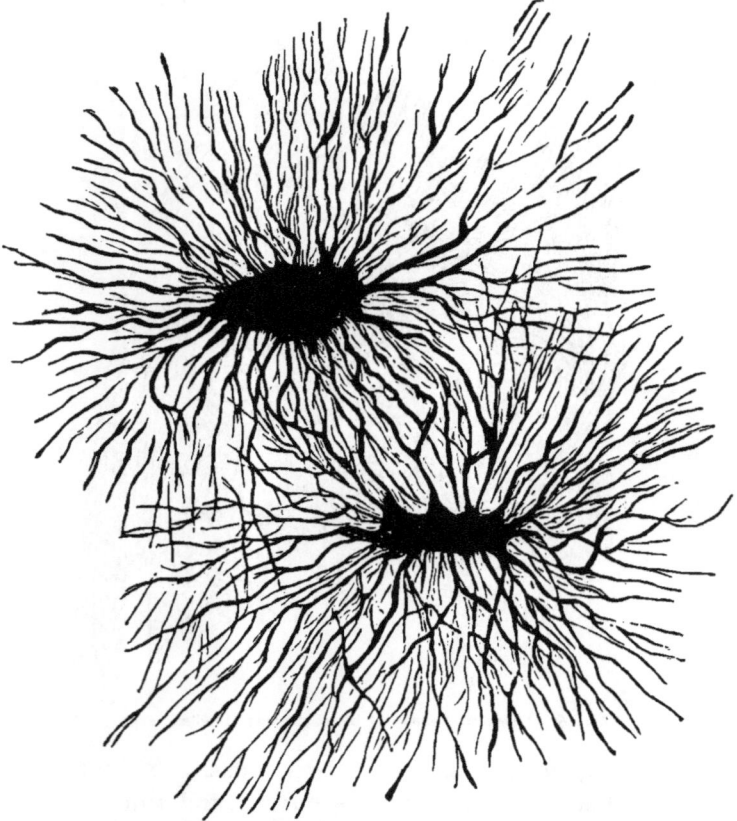

FIG. 47. Two bone corpuscles. It will be noticed that many of these canaliculi are not
connected with the lacunæ. They belong to other lacunæ not shown in
the drawing. x 600

fibres given off on either side, canaliculi ; these communicate
with others from other cells and thus the lacunæ are connected together. The majority of the canaliculi are given

off at right angles to the lamellæ but some pass off in all directions. In thin longitudinal sections this is well seen as is also their inter-communications. The canaliculi open in the Haversian canals, on the surface of the bone, and in the large central medullary cavity. In living bone the lacunæ are completely filled with protoplasmic cellular matter,— containing a nucleus and sometimes a nucleolus, — which sends prolongations into the canaliculi. Between the lamellæ of compact bone are fibrous bundles impregnated with lime salts and known as the "perforating fibres of Sharpey." They are connected with the periosteum from which they have their origin, and they are present in all bone developed in connection with the periosteum.

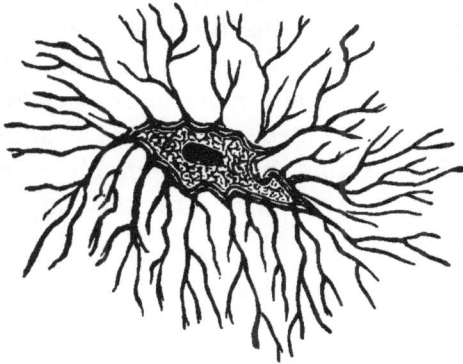

FIG. 48. Bone cells from fresh young bone of human tibia, filled with nucleated germinal matter. x 600.

This leads us to a consideration of the membrane surrounding bone, the periosteum. This is composed of two layers.

The external layer consists of a dense, firm, fibrous tissue. In some parts it covers the bone with but a single layer of connective tissue bundles, while in others two or three layers are recognized.

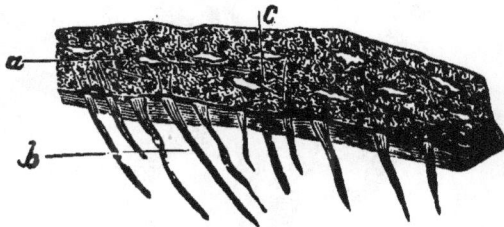

FIG. 49. The Sharpey's fibres, b, of a periosteal lamella of the human tibia. a, c, lacunæ. (from Frey.)

A few blood-vessels supply this layer. The internal layer has, in addition to the connective tissue bundles, a large number of nucleated cells of various sizes. In young growing bone, sharp points project into this layer, which are often covered with a layer of nucleated cells, " osteoblasts."

In the central cavity of long bones and filling them completely is a peculiar yellow or red substance called marrow. Both these varieties have a large supply of blood-vessels. Placed under the microscope the yellow marrow in seen to consist largely of fat cells, connective-tissue cells and nucleated cells similar in appearance to lymph corpuscles; these are the marrow cells. The red marrow as found in the meshes of spongy bone contains fewer fat cells and more marrow cells. Here are found large, colored, nucleated corpuscles. These are probably the intermediate forms between lymph cells and colored blood corpuscles. Here also are the large, many nucleated, giant cells of Robin.

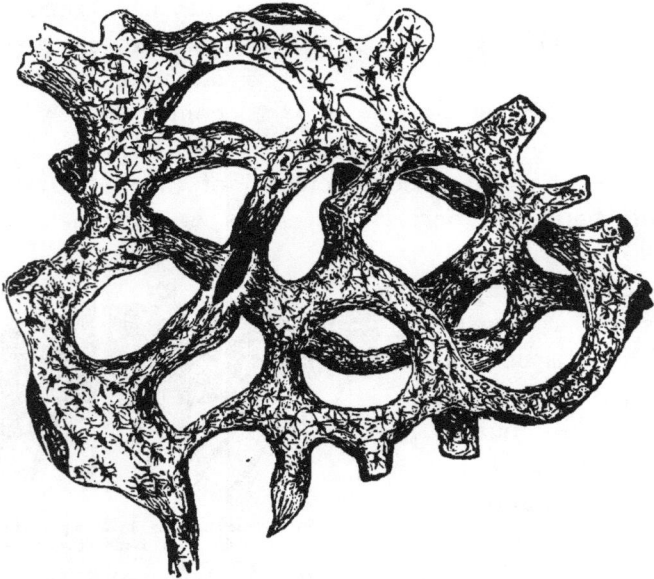

FIG. 50. Cancellated bone, from head of human femur. x25.

METHODS OF EXAMINING.

For a typical specimen of compact tissue a longitudinal and transverse section of one of the long bones of the body should be prepared, as for instance, sections from the humeras or femur. The bone selected should be entirely free from grease and of a pure white color. With a fine saw a thin section is cut and transferred to a hone or fine grindstone. In this laboratory oil stones are used, each stone measuring six inches long, two wide and one thick. The surfaces of these hones are freed from grease or dirt by washing them in warm soda water. The section of bone is placed between the flat surfaces of two of these hones. They are kept constantly wet with water. The lower hone rests on a table, while the operator moves the upper hone over it rapidly, and pressing hard upon it at the same time. The more force used the sooner will the work be over. These active measures are continued until the bone is as thin as a sheet of writing paper. Now the force is very slight, just the weight of the hone, and the motion is slower and slower. The section is ground in this way until it is extremely thin, as easily bent as thin paper, perfectly transparent, so much so that when wet and placed over the finger, it can scarcely be seen, so thin that under the ¼ inch only one layer of cells can be made out. This whole process need not occupy over twenty or thirty minutes. The section is now transferred to a glass slide and thoroughly washed in distilled water by aid of a camel's hair brush. It is then removed to a clean, dry slide and allowed to dry. If in drying it tends to curl, another slide had better be placed over it. The Canada balsam used should be hard when cold,—so hard that it can be chipped off in flakes with a knife. Ordinary balsam can be made in this condition by exposure to the air for a long time, or better still, by the application of heat for a short time, until the volatile matters are driven off.

A drop of this balsam, melted, is placed on a warm glass slide, also a drop in the centre of a cover glass. The slide and cover are kept warm over a flame until the balsam has evenly diffused itself. If any air bubbles appear they may be removed by touching them with a hot needle or by skimming them off by drawing the needle horizontally over the surface of the drop. The slide is now removed to the table and in a moment or two the cover also. When the balsam on the slide is slightly cool, but before it is cold, the bone is placed upon it. The cover is now inverted over it, and pressed against the slide. In a few moments the balsam is very hard, when the excess can be chipped off with a knife and cleaned by rubbing with a cloth moistened in turpentine. If the balsam be too hot when the bone is placed upon it, it will run into the lacunæ and canaliculi, and the specimen will not show to advantage; if too cold the cover glass can not be pressed down tightly. In this latter case the specimen can be gently warmed and the cover pressed down. The specimen is immediately exposed to the cold, to harden the balsam as soon as possible. If successfully treated nothing can exceed the beauty of these sections. The lacunæ and canaliculi are filled with air and surrounded by the balsam. They now appear intensely black and show to the best advantage possible. For a number of years we have followed this method and invariably have success. Over a thousand of these specimens are mounted by students in this laboratory each college year, and a poor specimen would be hard to find, notwithstanding students are recommended by some authors not to mount this tissue as it is so tedious to prepare and shows so poorly when prepared. After a little experience the whole preparing and mounting need not exceed thirty minutes. For longitudinal sections more care is necessary to keep the balsam from entering the cavities. It is advisable, therefore, at first not to grind these sections quite as thin as the transverse.

To macerate bone a .5 per cent. solution of chromic acid, to which has been added a few drops of hydrochloric acid, is employed. The bone should be cut in small pieces and the amount of the solution used should be very large. In a few days sections can be made in any direction with a razor. A saturated solution of picric acid, as recommended by Ranvier, is very useful. The pieces of bone should be small and crystals of the acid should be added from time to time. Fresh bones stained exhibit the bone cells, or the bone may be decalcified by chromic acid. Small pieces are immersed in a large quantity of the solution which should be very weak at first, 1 to 500, and gradually changed every day or two for stronger ones, until in a week it may reach 1 to 200.

The bone has been long enough in the mixture when a needle can be passed through its middle. Thin sections can be made with a razor ; after thoroughly washing in water to remove all traces of the acid, they are stained with hæmatoxylin and mounted in dilute glycerine. Decalcified bone is used to demonstrate Sharpey's fibres. One of the blades of a pair of forceps is inserted into the outer surface of the bone and a thin strip torn off. Examining several of these strips one will find on some the tapering fibres "looking like nails driven through a board." The flat bones of the skull are the best to use for this purpose.

CHAPTER VI.

Teeth.

A TOOTH may be said to be an enlarged papilla of the mouth which has undergone such histological and chemical changes that it has acquired a remarkable degree of hardness.

In the fully developed tooth there are three parts: 1, the crown, the free part projecting above the gums. 2, the neck,

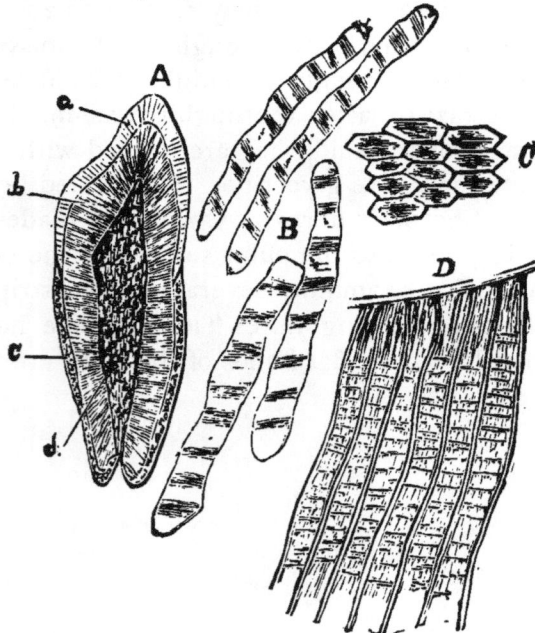

FIG. 51. A. Longitudinal section of a tooth. a, enamel. b, dentine. c, cementum. d, pulp cavity. B, enamel rods, isolated by acids, longitudinal view. C, transverse view of rods. D, the rods or prisms seen in situ. B. C. and D x 400.

surrounded by the gum. 3, the fang, the part projecting

into the alveolus of the jaw.   In the centre of the tooth is a canal with an opening at the apex of the root, terminating above after entering the crown.   It may be simple or multiple, depending on the number of fangs to the tooth.   The great mass of the tooth is composed of a substance much harder than bone termed dentine ; covering the crown of the tooth is the enamel, while surrounding the fang is a bony substance, cementum.

THE CENTRAL OR PULP CAVITY.

The central or pulp cavity is completely filled with a soft substance known as the dental pulp.   This consists of connective tissue, nucleated cells, blood-vessels and nerves. The nucleated cells are distributed through the mass of the pulp but mostly cover it as a distinct cell membrane.   They are oblong in shape and measure from $\frac{1}{1200}$ to $\frac{1}{800}$ of an inch in length and $\frac{1}{4000}$ in breadth.   This membrane, the membrana eboris, will cling to the walls of the pulp cavity when the pulp is removed.   The cells composing it were named by

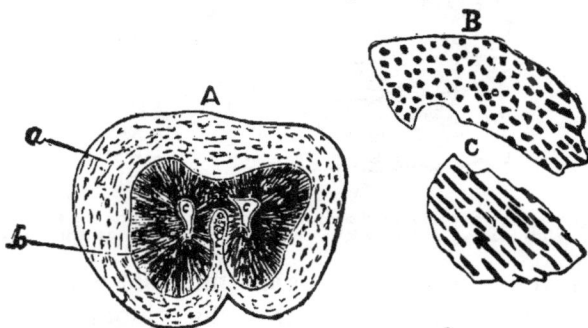

FIG. 52.   A, Transverse section of fang of a bicuspid tooth, x 5.   a, cementum.   b, dentine.   c, pulp cavities   B, transverse, and c, oblique sections of dentine.   x 400.

Waldeyer the odontoblasts.   Each cell sends one process or more into the tubules of the dentine, while other processes unite with those from neighboring cells in the membrane, and in the interior of the mass.   Thus all the deep and superficial cells are connected with each other and indirectly with the processes in the dental tubes. The vessels form a capillary net-work.

The nerves end in fine non-medullated fibres which are distributed at the surface of the pulp between the superficial cells.

Boll observed in the teeth of rodents, macerated one hour in a $\frac{1}{32}$ per cent. solution of chromic acid with the membrana eboris preserved in connection with the pulp, a large number of extremely fine fibres that passed outwards and, in teased preparations, accompanied the dentinal processes as fine hairs. By their length and direction they appeared to enter the dental tubes, and although no traces of them have been satis-factorily demonstrated, it is altogether probable that Boll's be-lief is the correct one.

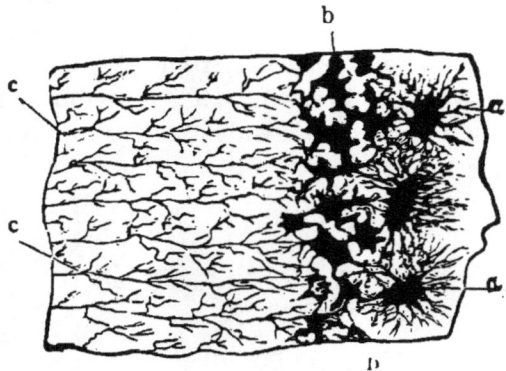

FIG. 53. Section through cementum and periphery of dentine. a, bone cells in cementum. b, interglobular spaces. c, fine dentinal tubes at surface of the dentine. x 400.

It is a well known physiological fact that nerves are more sensitive at their terminal points than along their course. Dentists find upon operating on the teeth that on reaching and cutting the periphery of the dentine great pain is experienced by the patient, but as soon as the cutting is deeper the pain is materially lessened. This is exactly what would follow if the nerve fibres terminated at the ends of the dental tubes, at the periphery of the dentine. If single nerve fibrils do not thus extend into these tubuli, then it would appear they must be connected in such a way with the odontoblasts that the processes of the latter are capable of transmitting the properties of the former. We are most favorably inclined to accept the views of Boll.

THE DENTINE.

The dentine presents a yellowish-white fibrous appearance and is one of the hardest constituents of the body. It has a firm matrix and extremely fine canaliculi, the so-called, dentinal tubes. At the larger ends their average diameter is about $\frac{1}{4800}$ of an inch and they are separated from each other by two or three times their width of matrix substance. They commence by circular openings on the walls of the pulp cavity and extend radially outwards, making spiral turns, like a corkscrew, on their way; this twisted appearance is easily seen in decalcified specimens. While extending outwardly to the enamel or cement they give off numerous branches by which they and their contents anastomose freely. At the surface of the dentine these are extremely fine and many of them terminate in larger or smaller cavities at this point known as the interglobular spaces of Czermak.

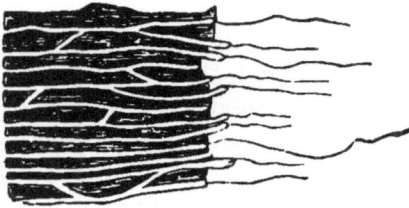

FIG. 54. Showing membrane lining dentinal tubules. (Boll).

The tubes are lined with a sheath, the dentinal sheath, which is readily seen in softened specimens. In these sheaths lie the dentinal fibres of Tomes. They are the greatly elongated processes of the odontoblasts with perhaps nerve elements. They are solid and homogeneous, and easily stained with carmine. In old teeth these fibres evidently do not extend to the finest terminal points of the tubes, although in the young they certainly do.

In that part of the dentine which is just beneath the cementum and sometimes in that just beneath the enamel, there is a large number of spaces, interglobular spaces, the granular layer of Purkinje.

Many of the dentinal tubes end in these spaces. They have a ragged outline and many short pointed processes. They denote an arrest in the development of the tissue at that point. Dental tubes pass through them uninterrupted (Tomes). Besides the tubes there is a substance within them which takes carmine staining only with difficulty. In a transverse section of dentine rings are seen concentric with the cavity of the pulp. These rings may be due either to curves of the dental tubes, each tube curving at the same distance from the surface or to rows of interglobular spaces.

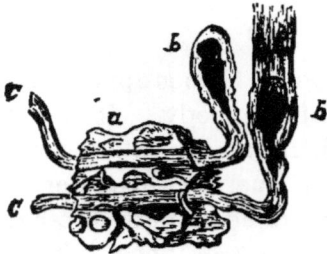

FIG. 55. Odontoblasts. a, portion of dentine. b, two odontoblasts which pass with their processes through a portion of the dentinal canals and protrude from them at c. (Beale).

CEMENTUM.

Cementum is absent from the crown of the teeth of man. (See cuticle of the enamel.) It commences just over the enamel at the neck of the tooth and forms a thick coating over the fangs. At the end of the root it is often found thickened by an exostosis. It is composed of a matrix identical with bone, and of lacunæ and canaliculi. The latter are much longer and more numerous than in true bone. They communicate directly with some of the dental tubes and also with each other. Some lacunæ are seen with sharply defined contours and with short processes. They are the "encapsuled lacunæ" first described by Gerber. In this way a single lacuna or several of them, may be enclosed. Here in the cementum, as in bone, are the penetrating fibres of Sharpey, representing calcified bundles of connective tissue. Where the cementum commences at the neck of the tooth no lacunæ are found and it appears structureless.

THE ENAMEL.

Upon the outer surface of the dentine of the crown of the tooth is the enamel, the hardest substance of the body. It is composed of closely crowded polyhedral prisms, the enamel prisms, enamel columns, enamel rods. They are about $\frac{1}{3000}$ of an inch in diameter and mostly pursue a direction from the dentine toward the surface. They are in close contact with each other, and so far as can be demonstrated there is no intervening substance to unite them together, although the action of certain reagents in isolating the rods leads one to suspect here, as elsewhere, a "cement substance." Nearly all the fibres run the whole length of the enamel but some are seen in the outer portions which do not penetrate far into the interior. Transverse lines or striations are seen on isolated fibres as well as continuous over adjacent ones. Hertz believes that these lines represent an "intermittent calcification" of the fibre. Tomes and Waldeyer think that they are due to varicosities in the individual fibre. It is a very remarkable appearance and difficult to account for. If hydrochloric acid be added to the fibres after they have been isolated, they will break up into small cubic fragments of about equal size, corresponding to the striations on their surface. With the exception of these striæ, the enamel rods appear perfectly homogenous, yet it is observed that acids act upon the central part of the fibre, before they do on the periphery. This is readily understood when the formation of the enamel is understood. The hardening salts are deposited first in the periphery of the cells and gradually reach the centre, so that in immature fibres may be seen a central canal. Soon this difference is obliterated as calcification progresses ; but when the acids act upon them this action is reversed, the more recently deposited calcified substance is sooner affected.

There are coarser striations, consisting of a series of concentric lines crossing the enamel fibres. They are of a brownish color and are known as the "brown striæ of Retzius." It is possible that they mark the different stages of the growth of this structure. While one end of the fibre is implanted in depressions in the dentinal surface, the other terminates as a free end to form part of a beautiful hexagonal mosaic.

THE CUTICLE.

Covering the surface of the enamel is an exceedingly tough membrane, the cuticle of the enamel, Nasmyth's membrane. In thickness it is not more than $\frac{1}{5000}$ to $\frac{1}{25000}$ of an inch. In young teeth this is easily detached after slight action of hydrochloric acid, but it is doubtful if it exists in the teeth of the adult. Although very tough and unaffected by acids, yet it is not so hard as the enamel and is on this account generally worn away. Silver staining shows it to be composed of cells of an epithelial type. On its under side are the indentations for the reception of the free ends of the enamel fibres. Tomes regards this cuticle as a thin covering of young and incomplete cementum.

METHODS OF EXAMINING.

Sections of unsoftened teeth can be made in any desired direction, ground and mounted precisely as recommended for bone, giving the very best results. The methods for softening teeth are the same as those for softening bone. A 10 per cent. solution of hydrochloric acid is generally useful. If this solution be strengthened the dentinal substance will be destroyed and the sheaths lining the tubes will remain for a considerable length of time. Only young developing teeth in a fresh condition should be subjected to the action of acids for the purpose of isolating the enamel rods. Their transverse lines may be seen by adding muriatic acid.

The dental pulp is studied.by using a reagent that softens the parts around it, at the same time that it hardens the pulp itself. Such a reagent is picric acid. The fresh tooth is broken open by the blow of a hammer and placed at once in a saturated solution of the picric acid. More crystals of the acid are added from time to time and the tooth is frequently stirred in the mixture in order that fresh parts of the solution may come in contact with it. As soon as the tooth is soft enough to allow a needle to pass through it, it is transferred to alcohol. The alcohol is changed daily until it fails to be colored by the acid. Thin verticle sections are cut with a razor, stained n hæmatoxylin, and mounted in glycerine.

CHAPTER VII.

Muscle.

MUSCLE may be divided into two general classes:

1. Striated,	1. Non-striated,
2. Striped,	2. Smooth,
3. Fibres of animal life,	3. Fibres of organic life,
4. Voluntary,	4. Involuntary,
5. Responds rapidly.	5. Responds slowly.

Thus there are five different terms applied to each of the classes. Some of them are based upon histological and others upon physiological distinctions.

There is one muscular organ in the body that cannot be classed with either of these divisions. It not only possesses properties belonging to both, but has in addition characteristics not found in other muscles. The heart deserves a place by itself and will be treated apart from striated muscle, although its intimate structure is so nearly identical with it. In studying striated muscle the unaided eye at once discovers a thin membrane surrounding the whole muscle and sending prolongations into the body, giving the familiar appearance of a fine or coarse grained muscle. If a muscle be cut transversely this membrane will show to good advantage. The external investing membrane consists of a more or less dense connective tissue known as the perimysium, while the portion running through the muscle, dividing it into compartments, is called the endomysium. Each of these compartments is a fasciculus, and a fasciculus is a bundle of small fibres, the cut ends of which are seen in the figure as small dots. These fibres may extend through the whole length of the shorter muscles, but usually in the skeletal muscles they are only from 1½ to 2

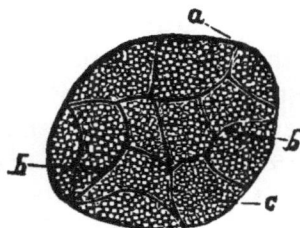

FIG. 56.   Transverse section of a small muscle of a frog. a, perimysium. b, endomysium.   c. cut ends of muscle fibres, showing as dots. x 15.

inches in length, although this varies greatly in the same muscle and in different muscles.   The diameter of the fibres is also of varying size from the thickness of a single contractile disc, to the $\frac{1}{250}$ of an inch in man, and to a very much larger size in some of the lower animals.   Their diameter will average not far from the $\frac{1}{500}$ of an inch.

This diameter is said to be much less in the female, but practical experience will not warrant the assertion.   These fibres end rather abruptly, the sarcolemma extending over the end of each fibre and becoming lost in the inter-fibrillar connective tissue.   The sarcolemma is a transparent, homogeneous, very thin and highly tenacious membrane closely investing each muscle fibre.   It is invisible in fresh muscle, but is easily demonstrated by the aid of reagents.   Figure 57 shows this membrane extending from one end of the broken fibre to the other.   To obtain this view a small piece of frog's muscle was teased and while looking through a dissecting microscope, pressure was made with a needle over one

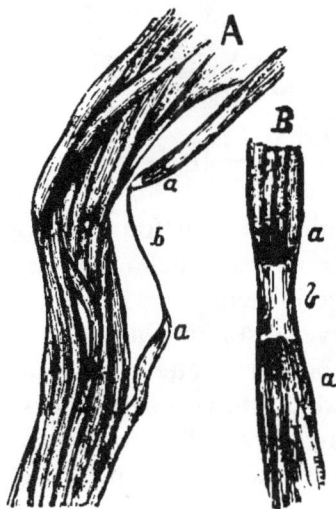

FIG. 57.   Sarcolemma of muscle (frog's). A. a, ends of a broken fibre. b, sarcolemma. x 35.   B, Showing sarcolemma, more highly magnified.

of the fibres.   The muscle

substance was broken and then contracted either way, leaving the sarcolemma intact.   By staining with carmine the membrane was sufficiently colored to enable us to procure good micro-photographs of the specimen.   Just beneath the sarco-lemma are the numerous nuclei of the muscle fibre.   They are situated between the muscle substance and the sarcolemma, and do not form a part of the latter, neither is the sarcolemma developed from these nuclei, but from others, no trace of which can be seen in the adult muscle.   According to Klein each nucleus contains an intra-nuclear fibrillar net-work.   They are   much   more   numerous   in young,   growing   muscle.   Rollett describes   the   nuclei   found   in muscles   of   the   amphibia,   fishes

FIG. 58.   Striated muscle fibre of the frog.   The nuclei stained with hæmatoxy-lin.   x 200.

and birds as existing within the muscle   substance,   a   condition similar to that found in the heart of man.

These nuclei are easily demonstrated by adding dilute acetic acid to a fresh specimen, or by employing one of the several staining fluids.   They are often seen surrounded with a border of finely granular substance.   A muscle fibre is divided into two substances by broad dim bands and bright narrow ones.   The former is the contractile part of the fibre and is composed of contractile discs, while the narrow bright bands correspond to the interstitial discs.   During contraction the first become more transparent, thinner in their longitudinal direction and correspondingly thicker transversely.   The second become more opaque.   At figure 59 a, is seen a contractile disc, while b. represents the interstitial disc.   Thus in living muscle it is seen there are no longitudinal striæ. By studying one of these contractile discs more carefully it becomes differentiated into thin oblong rods, each rod the length of the disc. Each of these rods represents a single sarcous element, which is the anatomical element of the contractile disc.

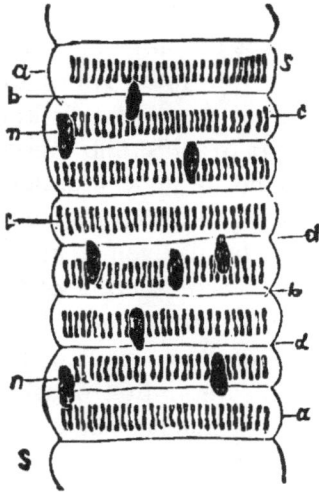

FIG. 59. Muscle fibre. a, contractile disc. b, interstitial disc. c, sarcous elements. d, transverse membrane of Krause. n, nuclei. s, sarcolemma, (Klein.)

These sarcous elements are arranged so close to one another during the life of the muscle, that they appear as one, and the disc was said to be homogeneous. However, by the use of reagents, or many times spontaneously, during and after life they become separated when a fluid substance is pressed out which áppears to be identical with the myosin discovered by Kühne. Now if alcohol be used for hardening the muscle, the sarcous elements will be arranged endwise, and as a result of these elements being placed end to end, we have the appearance of long slender fibrils, the primitive fibrillæ. If, however, in the place of the alcohol, hydrochloric acid be used, then the sarcous elements appear arranged sideways and we have the transverse discs. A muscle fibre then is either a bundle of primitive fibrils or of transverse discs which ever way we look upon it.

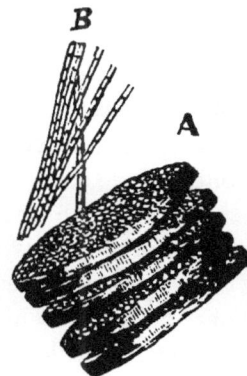

FIG. 60. Muscle fibre of frog. Separating into A, transverse discs. B, longitudinal fibrils. Prepared from separate specimens. x 300.

Very intimately connected with the sarcolemma of a fibre,

are membranous septa which stretch across the muscle at regular intervals.   These septa are the transverse membranes of
Krause.   They divide the fibre into a number of equal sized
compartments, "known as muscle compartments." This membrane of Krause passes directly across the fibre midway between two contractile discs, dividing the intervening interstitial disc into two equal parts.   Each of these parts is known
as a lateral disc, each disc belonging to different muscular
compartments.   A small granule is found in some muscles directly at the end of each sarcous element.   This arrangement is
so constant in some muscles, that it has been given a name, the
"granular layer of Flögel."   A transverse section of these
sarcous elements presents a fine granular appearance, leading
one to believe that they are composed of most minute fibrils.
It is certain that they are not optical units, but consist of
minute doubly refractive elements, the "discliaclasts of Brücke."
    A muscle then is a collection of fasciculi.
    A fasciculus is a collection of fibres.
    A fibre is a collection of muscle compartments.

A muscle compartment is a collection of { Transverse membranes of Krause. Sarcous elements. Myosin, and sometimes the granular layers of Flögel.

This kind of muscle is found in all the skeletal muscles
of the body, in the muscles of the oral cavity, pharynx, larynx,
œsophagus, lower part of the rectum, diaphragm, middle and
outer ear, sphincter vesicæ, part of muscles of the prostate, and
modified in the heart.   Although a few of the striated muscle
fibres of the body divide, yet such is not the rule.   Aside from
these few muscles the heart presents distinctive characters,
particulars in which it differs histologically from the ordinary
striated muscles described above.   First, the fibres have no
sarcolemma; second, they are smaller; third, they divide; fourth,
they anastomose; fifth, they are divided into nucleated cells;
sixth, the nucleus is within the muscle substance of each cell.

Figure 61 represents some of these peculiarities. The fibres are much smaller in some parts of the heart than in others. In examining a transverse section some fibres will be found very small indeed. These small fibres are the ends of one of the divided fibres, and when it is remembered that they frequently divide and terminate in a pointed end, this great irregularity in size can be readily accounted for. At quite regular distances the fibres are crossed with a faint line, and midway between two of these lines is a nucleus which is situated within the substance in the centre of the cell. Many times these cells are so arranged that on a thin section, it appears that the end of one cell is placed just opposite the centre of another near it, so that the ends of the cells look not unlike a series of steps.

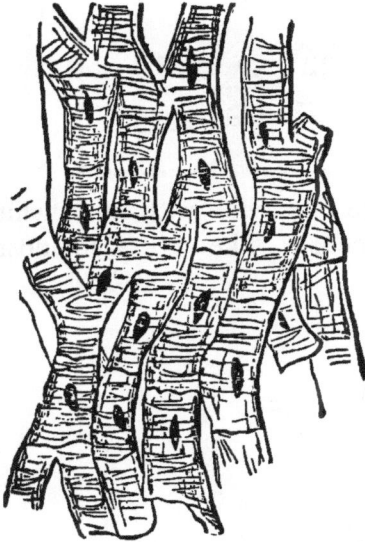

FIG. 61. Muscle from the heart of man, prepared from fresh specimen, cut with freezing microtome and stained. x 150.

Meyer tells us that the more a muscle works the deeper is its color, but according to Ranvier there may be well defined structural differences between the pale and the deeply colored muscles.

If the diaphragm be examined a great number of large nuclei or "muscle cells" will be found. They may be in sufficient numbers to form a nearly complete layer around the fibre. All fibres do not seem to be affected alike in this respect. The work of the diaphragm must necessitate a great amount of waste and repair, hence the large number of bioplasts.

The second class, involuntary, unstriped muscle, is widely

diffused throughout the body. (See Stricker, pp. 150, 151.)
This tissue is aggregated into larger or smaller bundles and is
composed of elongated spindle-shaped cells held together by a

FIG. 62.   Muscle from diaphragm with large number of bioplasts. (Klein.)

transparent semi-fluid substance identical with that which
unites epithelial and endothelial cells.   The cell is composed

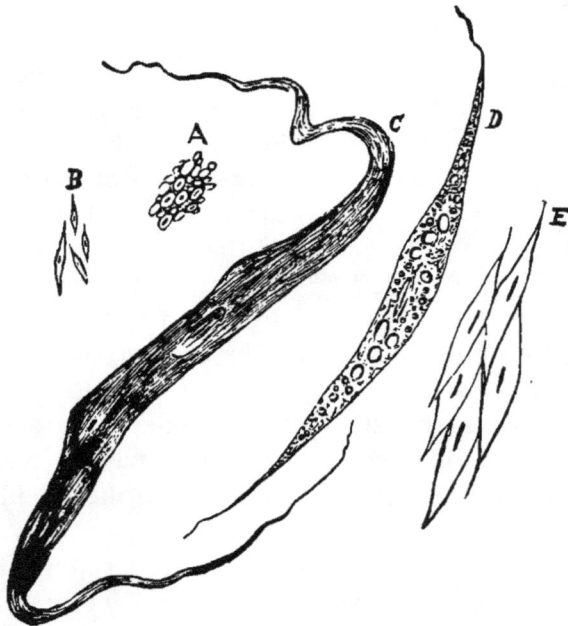

FIG. 63.   Muscle cells.   A, transverse sections through bundles of smooth fibres.   B,
cells from small artery of guinea-pig.   C, cell from intestine of man.   D, cell
containing fat globules (uterus).   E, cells from artery (man).   x 400.

of a longitudinally striated substance in the centre of which is
a nucleus not infrequently multiple, and usually in the middle

of the long axis of the cell, or in the thickest part of it, which
may bring the nucleus nearer one end than the other.   The
nucleus is usually oblong and is composed of a fine net-work
of fibrils which anastomose (Klein) at the poles of the nucleus
with the bundle of fibrils composing the central part of the
cell, the true contractile substance.

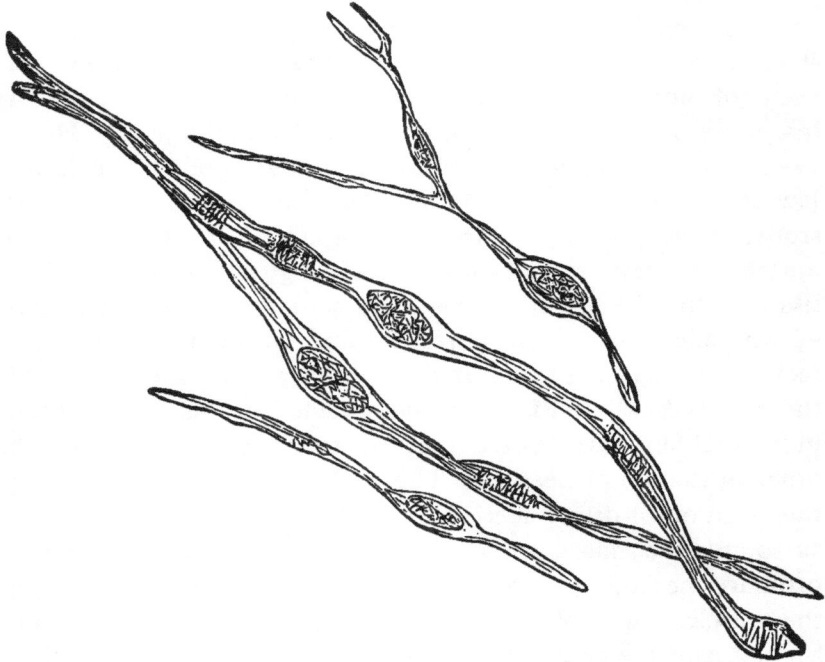

FIG. 64.   Non-striated muscle cells, (after Klein.)

Each cell is surrounded by a fine sheath, which shows a
transverse linear marking, especially if the cell be examined in
a contracted state.   The ends of the cell are drawn out into
fine points, and in the arteries and veins the extremities are
frequently branched.   The longest and thickest cells are found
in the walls of the intestine, the shortest in the arteries, the

thinnest in the tubes of the sweat glands (Klein.)   Fat granules are frequently imbedded in the cells.   This is well seen if
some of the cells be examined from the uterus a few weeks
after delivery.   While some will be filled with fat, others will
be nearly destroyed by the degenerative process.   Granules
are nearly always present at the two ends of the nucleus.

METHODS OF EXAMINING.

To examine microscopically the general appearance of
striated muscle, a small fasciculus may be taken from the
body of any of the vertebrates, and by a slight amount of
teasing in some normal fluid, good views are obtained.   However, it is better to remove from the under side of the lower
jaw of the frog one of the thin flat muscles so suitable for
study.   The frog should first be decapitated or pithed and the
muscle removed carefully without teasing or straining.   The
tissue should be placed on the glass slide at once and moistened with blood-serum or normal saline solution.   Very satisfactory views of the transverse striations and in some cases of
the longitudinal striations will be obtained in this way.   If the
individual fibres are desired, they are easily separated from each
other by the aid of needles.   The nuclei can be recognized after
the addition of dilute acetic acid.   The sarcolemma so closely
surrounds the muscular elements that it is not visible by the
ordinary methods of examination.   Yet many times in teasing
the muscle, some of the fibres have been pressed upon and
broken, and the contractile substance has contracted at either
end, leaving the clear transparent sarcolemma as seen in
figure 57.   If water be added to a fresh specimen it will soon
pass through the delicate sarcolemma, causing it to separate
from the tissue, so that there is a transparent border at the
edge of the fibre.   Many times, too, the sarcolemma will bulge
out at the end of a fibre in the form of a little pouch.   To
study farther, some of the tissue may be placed for a few days
in a .2 per cent. solution of hydrochloric acid, then the sarcous

elements will appear arranged sidewise into transverse discs, each disc equalling in length a single sarcous element, and in width the same as the muscle fibre.   At the same time place a specimen in a .5 per cent. solution of chromic acid.   If some of the tissue be placed in Müller's fluid for a couple of weeks or longer, but slight teasing will cause the muscle fibrils to fall apart. Alcohol and chromic acid will cause the sarcous elements to become arranged longitudinally, giving the appearance of longitudinal fibrils, each fibril being just the width of a single sarcous element, and as long as the muscle fibre itself.   To demonstrate the relation of muscle to tendon, a small shred of muscle with its tendonous attachments must be placed from 20 to 30 minutes in a 35 p. c. solution of potash, when the appearance seen at figure 65 will frequently be observed. It will be noticed that the sarcolemma is still intact, but that the fibres of the tendon have become separated from the muscle fibre. There has been no teasing or injuring of the tissues, and one is forced to believe that the two were cemented together, and that the potash solution dissolved this cement.   It seems safe to assert that muscle is united to tendon by a cement, which is dissolved by a 35 p. c. solution of potash in from 15 to 30 minutes.   The examination of muscle for trichinæ is very simple.   Small shreds of the muscle should be teased with needles in some normal fluid media, and examined at once with a low power; one giving 50 diameters will be sufficient at first, although for a more careful examination one of 250 or 300 diameters should be employed.   Thin sections can

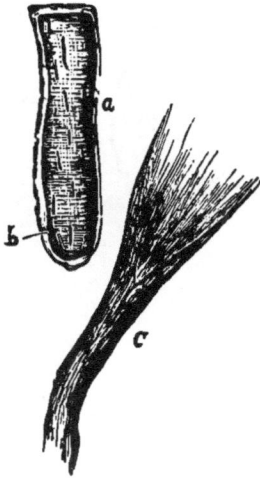

FIG. 65.   Termination of muscle in tendon, from tail of young mouse.   a, muscle fibre.   b, sarcolemma.   c. tendon.   x 50.

be made with a razor through the trichinous muscle hardened in alcohol, using the proper care that the sections be made in the direction of the fibres.   The worms are seen coiled up as

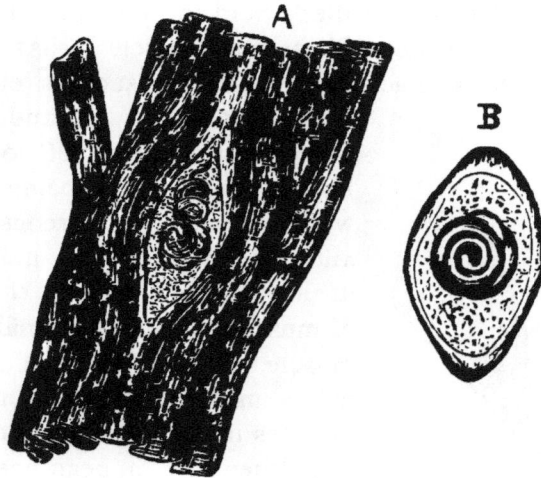

FIG. 66.   Trichinous muscle.   A, from psoas muscle of hog.   B, encysted trichina from arm of man.   B, x 35.

in figure 66.   They may be found in any of the transversely striated muscles with the exception of the heart.   They are most frequently found, however, in the diaphragm and muscles of the jaw and neck.   They are in greatest abundance at the tendonous extremities of the muscles, for they are here prevented from moving farther.   In the hog, fragments of muscle should be examined especially from the ham and tenderloin. They are usually found arranged spirally just beneath the sarcolemma.   This spiral arrangement gives them their specific name, trichina spiralis.   They were discovered in January, 1860, by Professor Zenker of Dresden.   In 1864 Professor Dalton counted the number of trichinæ in a piece of muscle $\frac{1}{12}$ inch square and $\frac{1}{30}$ inch thick, and found 12.   This would

give about 85,000 to the cubic inch.   In another specimen of
the same size he found 29 trichinæ, giving again in round
numbers 208,000 to the cubic inch.   The trichinæ in half a
pound of infested meat would be sufficient in a few days to
develop the extraordinary number of 30,000,000.   When it is
remembered that each of these worms must puncture the
mucous membrane in its way to the muscles, it is readily understood why they should occasion such notable disturbance.
As found in muscles they are usually surrounded by a cyst
containing granules or calcareous matter.   In size they aver-
age about $\frac{1}{30}$ inch in length and $\frac{1}{800}$ inch in thickness.   They
retain their vitality in the encysted state for a great length of
time.

Muscle sometimes becomes streaked with fat when it can
be examined either fresh or after being treated with Müller's

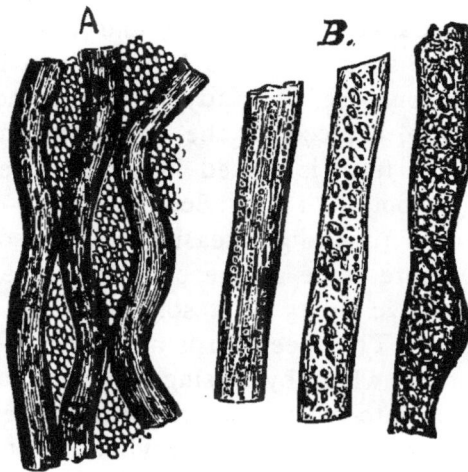

FIG. 67.   A, fatty infiltration of heart (man).   x 75.   B, fatty degeneration of muscle
from arm of boy, amputated on account of paralysis of three years standing.

fluid or chromic acid.   When the fat is confined to the connective tissue between the muscle fibres as in obesity, nothing
serious can result from it, unless in some particular organs·

when it may cause a dilatation and weakening from pressure. A specimen of fatty infiltration is seen at figure 67.

When the fat molecules are arranged in rows corresponding to the longitudinal striæ, they usually increase in number until all the contractile substance disappears and the muscle fibre becomes transformed into a row of fat cells, the fatty degeneration causing a complete wasting of the muscle. Nonstriated muscle cells undergoing this fatty metamorphosis are found in the uterus a few weeks after delivery, when in order for this organ to become reduced to its previous size many of the cells become thus broken down and carried out of the body. Striated muscle is best preserved in dilute glycerine, equal parts of glycerine and water,—which has been slightly acidulated by acetic acid. The glycerine will have the effect of causing the transverse striæ to show exceedingly well, aside from being a most excellent preservative fluid. Muscle may be preserved in Canada balsam also, and always should be so mounted if the specimen is to be examined by polarized light. For the purpose of studying the spindle cells of involuntary muscle, a piece of the organ containing it, as the bladder of the frog, is placed in a very weak solution of bichromate of potash (1 to 800) for forty-eight hours. At the end of this time slight teasing with needles will be sufficient to separate some of the cells from the large piece. Dilute nitric acid, 20 or 25 p. c. solution, does not fail to give good results. The specimen is allowed to remain in the acid 30 to 36 hours, when, by teasing in water each cell can be brought clearly into view. Before teasing, the tissue may be stained to advantage with hæmatoxylin. By removing the epithelial cells from the inner surface of the bladder of the frog, by washing and brushing with a camel's hair brush, and by staining with picrocarmine or hæmatoxylin, most beautiful specimens can be obtained. A specimen mounted in glycerine and treated as just described, is in the author's possession, showing the nuclei of the cells, the stained blood-vessels, and the blood-corpuscles with stained nuclei in the vessels. Nitrate of silver affords the best exhibition of the arrangements of the cells.   About .5 per cent. solution is allowed to come in contact with the tissue for two or three minutes only.   This is washed off with distilled water and the specimen is placed in dilute alcohol and exposed to the direct rays of the sun.   In a few minutes it can be mounted in glycerine, when the outline of each cell will be distinctly seen.   A one or two p. c. solution of acetic acid will show quite distinctly the individual cells, only a few minutes immersion being required.   Weak solutions of chromate of ammonia are recommended by Klein.

CHAPTER VIII.

Blood-Vessels.

THE blood is conducted from the heart by highly elastic arteries with a circular muscular element, which relatively increases in thickness as the arteries diminish in size. It is returned to the heart by the less elastic veins which have a variable muscular element. Between the two systems of vessels are the capillaries of exceedingly small but variable size.

FIG. 68. Capillary from mesenterium Treated with nitrate of silver and stained with hæmatoxylin. a, cell plates. u, nuclei. s, stomata. x 400.

THE CAPILLARIES.

The finest capillaries in the body are barely sufficient to allow the passage of the blood-corpuscles, one after the other, often so small as to compress them and change their form. They may be said to vary from $\frac{1}{6000}$ to $\frac{1}{2000}$ of an inch in diameter. In the nervous tissue and retina they are found the smallest, in mucous membranes they are of medium size, while in the bones and glands they are the largest. They do not diminish or increase in size as do the arteries and the veins,

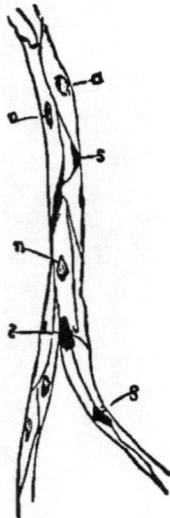

112

for they form a plexus of vessels of nearly uniform diameter, inosculating in every direction. Their average length is about $\frac{1}{50}$ of an inch. Although their diameter is so small and their length so short, yet the number is so large that their capacity is immense. It has been stated that the entire capacity of the capillary system is a number of hundred (400 to 800) times as great as that of the whole arterial system. These estimates, however, partake of the curious and are mere suppositions.

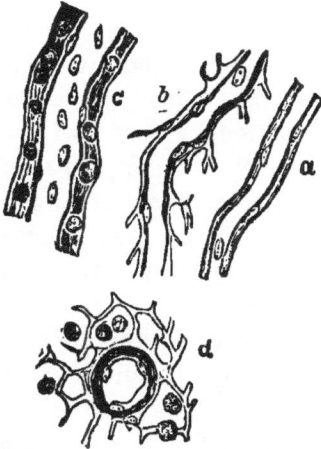

FIG. 69. Capillary vessels and fine branches of the mammalia. a, capillary vessel from the brain. b, from a lymphatic gland. c, a somewhat larger branch with a lymph-sheath from the small intestine, and d, a transverse section of a small artery of a lymphatic gland. (Frey.)

Without the aid of reagents the structure of the capillaries appears very simple. A few nuclei with nucleoli are seen scattered along the walls of an elastic hyaline membrane. Nitrate of silver dispels this illusion and resolves this homogeneous membrane into a single layer of nucleated cell-plates, united together by a "cement substance," which is stained a deep black by the solution.

If the capillaries become distended in any way, this cement is liable to give way and minute openings appear, which after a time may become enlarged into stomata. These openings may be present in the capillaries when they are not distended, but as a rule in this condition of the walls of the capillaries these "stomata" are not true openings, being covered with the cement substance.

Through these stomata the white corpuscles migrate and the red find an exit, being forced out passively. Purves, proved that the white corpuscles migrated through these openings, by staining capillaries through which emigration had been going on.   In some of the larger capillaries there is an outer sheath of connective-tissue cells surrounding this single layer of epithelial cell-plates.   This layer forms a reticulum, in the meshes of which are lymphoid elements.

THE ARTERIES.

Passing now to the larger vessels a gradual increase in diameter is observed, and transversely arranged nuclei are seen.   This is the very commencement of the muscular layer, and marks it as a commencing artery in contradistinction to a capillary.   If a medium sized artery be examined as directed below, four layers will be recognized.   The most internal endothelial layer consists of a layer of flattened nucleated endothelial cells, rendered visible by nitrate of silver staining.

By endothelium is understood a layer of flat cells which covers any membrane not a mucous membrane or one lining the cavity or canal of a secreting gland.   They

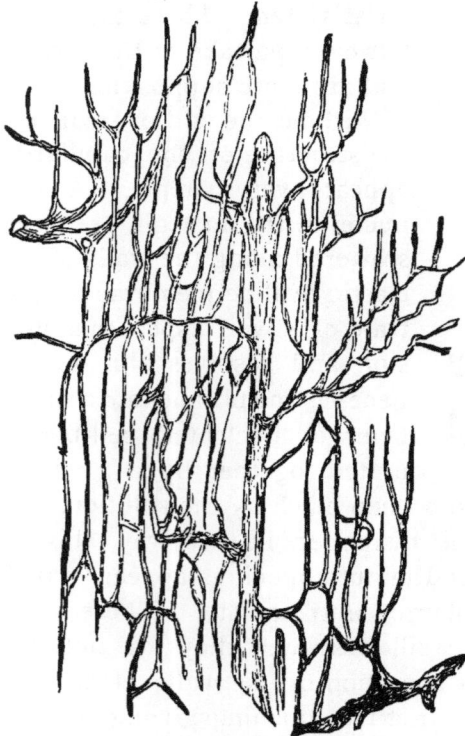

FIG. 70.   Capillaries of injected muscle of cat. x 100.

vary in their shape and outline, are held together by a "cement substance" and are demonstrated by the aid of a .25 to .5 p. c. solution of nitrate of silver.

External to this layer is a longitudinally striated membrane, the hyaline elastic coat. The net-work of elastic fibres is arranged parallel to the long axis of the vessel. The third coat from within, the intima, is termed the middle layer of unstriped muscle. The cells are arranged for the most part transversely. It is the most conspicuous of all the coats in all the arteries and is very distinct in the larger ones. The muscle cells first formed only a single layer, but they gradually increased until now they constitute an individual coat. In most of the arteries there is a fine granular connective tissue and a few elastic fibres between the muscle cells. Some of these cells are arranged obliquely, or even longitudinally. In some arteries there are no muscle cells whatever. The fourth, or most external coat, the adventitia, is composed of connective tissue with longitudinally arranged cells. Between this and the intima, in the larger arteries, there is an elastic membrane, the external elastic coat. It is composed of a network of fine elastic fibres. The adventitia is relatively very thick in small arteries, being as thick as the muscular layer. In large arteries it is not nearly as thick, and at the commencement of the aorta, it is a very thin plate.

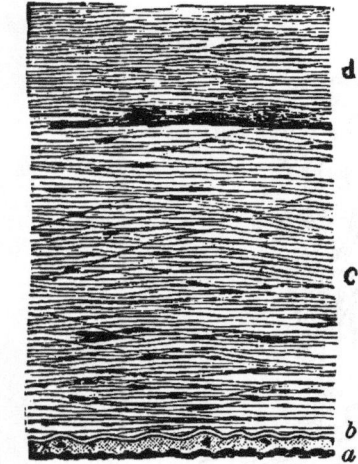

FIG. 71.   Transverse section of the walls of an artery. a, lining, endothelial layer. b, elastic layer. c. muscular layer. d, connective-tissue layer. x 100. (Quain.)

THE VEINS.

Commencing at the capillary there is observed, first, beside the endothelial layer, not a layer of muscle cells as in the

artery, but a layer of a fine longitudinal net-work of fibres.
The cells of the endothelial layer are shorter and broader, and
are not as fusiform in shape as the corresponding layer in
arteries.   The adventitia is about the same as that in the arteries.   In a large number of the veins, muscular tissue is
found in this external coat.   The whole walls are thinner than

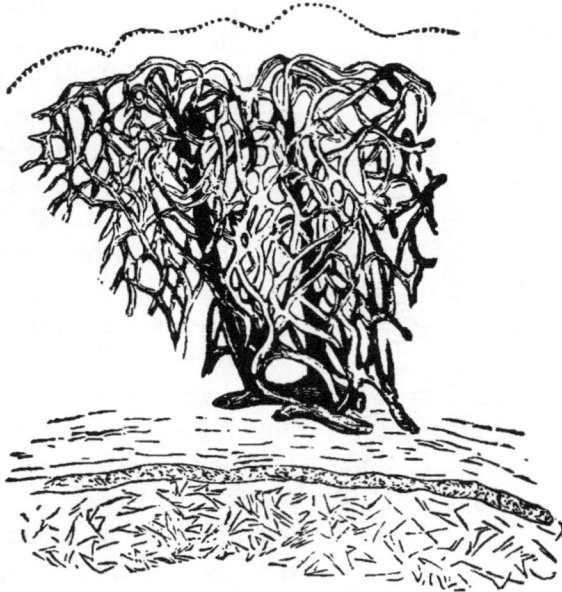

FIG 72.   Blood-vessels of the stomach of a cat.

in arteries of corresponding luminæ, and the yellow elastic tissue that gives to the large arteries their thick walls is so
scanty in the veins that when they are cut across they collapse
and their cavity is obliterated.   The valves of the veins are
folds of the intima and part of the muscular medium, although
some veins are entirely without this muscular medium.   Arn-
stein and Steida claim that the trunks of the venæ pulmonales
have striped muscle in their walls.   Steida says there is an inner circular and outer longitudinal coat continued from the

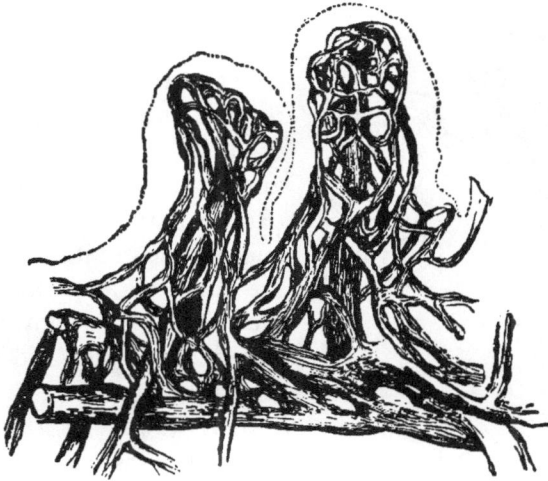

FIG. 73. Two villi from ileum of infant. x 100.

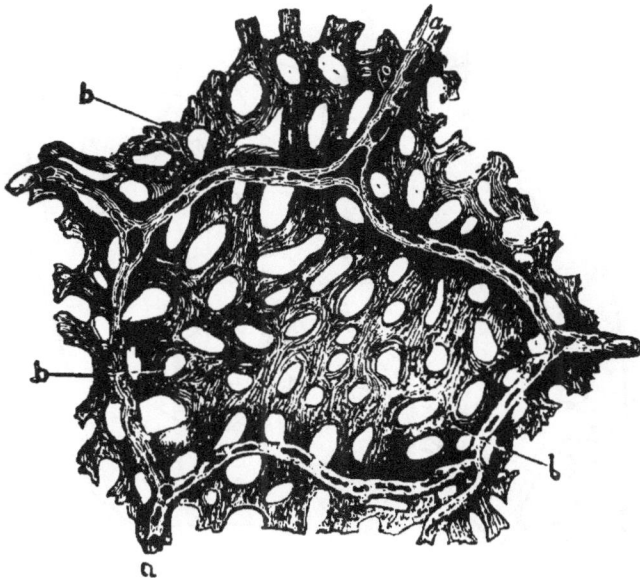

FIG. 74. Blood-vessels of human lung. a, larger vessels. b, capillary net-work. x 400.

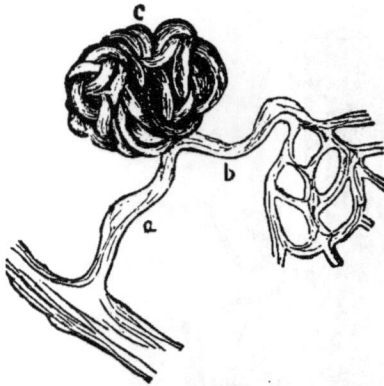

FIG. 75. Looped capillaries in kidney. a, afferent vessel. b, efferent vessel. c, glomerulus. x 75.

muscular walls of the left auricle. The arteries sometimes terminate in veins without the intervening capillaries, as in the tip of the nose, fingers and toes, and matrix of nails (Smith.) In the cavernous tissues of the genital organs there are large wide sinuous spaces, the walls of which are composed of elastic and unstriped muscular tissue. The arteries convey blood to these spaces and the veins carry it back.

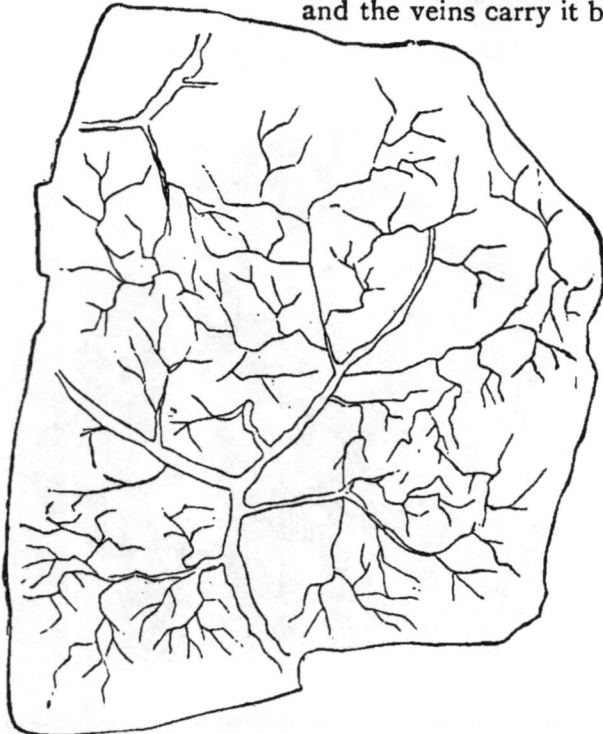

FIG. 76. Blood-vessel beneath mucous membrane of intestine of infant.

The walls of the blood-vessels are supplied with quite a rich plexus of capillary blood-vessels which penetrate the external layers, but never, the internal. They are called the " vasa-vasorum " and are in no wise connected with the vessels on which they are distributed. They arise from some vessel at a considerable distance from their point of distribution and supply the walls with nutritive materials.

Nerves are freely distributed.

METHODS OF EXAMINING.

The walls of capillaries can be studied best by isolating some of the vessels of the pia mater. A small piece of brain substance is clipped from the exterior of a con-volution with the pia mater attached. The section is removed to a slide with the pia mater on the glass. By aid of a camel's hair brush the cerebral matter is washed away and a drop of a .5 per cent. silver solution added. In two or three minutes this is removed and distilled water added. The specimen is exposed to the light until it is colored brown, when it is placed in glycerine and mounted. The capillaries in the mesentery of the frog, or tail of the tadpole, can be demonstrated in this way. The silver maps out the cells with unerring accuracy, and encloses a nucleus in each one. In the vessels from

FIG. 77. View obtained by changing focus of figure 76, showing villi of intestine, etc. x 35.

the pia mater, one frequently sees the termination of a capillary and commencement of an artery, by the appearance here and there of a muscle cell coiled round the vessel. With a freezing microtome transverse sections of the different arteries and veins can be made from the fresh specimens.

The silver method is recommended for the study of the endothelial layers of all vessels.

For the intima, a section of the vessel is placed in a 1 per cent. solution of potassic bichromate for several days, then it can be pulled off with forceps, stained, examined, and mounted in glycerine. Chromic acid preparations are valuable here. For the further study of this layer consult the chapter on muscular tissue.

CHAPTER IX.

The Respiratory Passages.

THE lungs first appear as an elevation on the anterior surface of the canal which becomes the œsophagus into which
the elevation soon opens, forming a hollow sac.   At its lower
extremity a bifurcation appears, dividing the tube into two

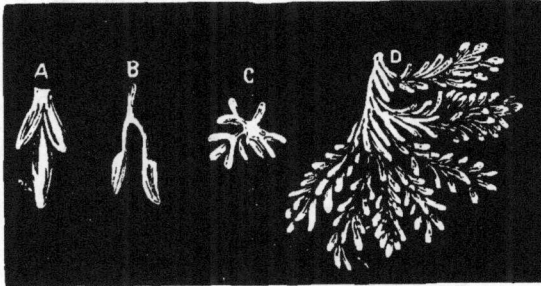

FIG. 78.   Formation of the bronchial ramifications and of the pulmonary cells.   A, B,
development of the lungs, after Rathke.   C, D, histological development of
the lungs after J. Müller.   (Longet)

branches.   These branches divide again and again until the
whole of the bronchial system is complete.   Then the pulmonary vesicles are developed and last of all the trachea.

THE LARYNX.

The general surface of the larynx is covered with a layer

of stratified epithelial cells, largely of the ciliated variety. Beneath this is a layer of mucous membrane containing a network of capillaries. Between these two layers can occasionally be demonstrated a layer of connective-tissue cells.   The nerves terminate in the mucosa in the form of terminal bulbs. The epiglottis is composed of a basis of reticular or elastic cartilage and deep layers of epithelial cells cover its anterior surface, while the posterior surface has a much less covering of the same kind.

THE TRACHEA.

The trachea is a fibrous tube, composed of fifteen or twenty hyaline cartilaginous half-rings embedded in its anterior walls, together with all the layers present in the larynx. On its internal surface is a continuation of the ciliated cells of the larynx. Just beneath this layer is a thick basement membrane. External to this is the mucosa, and still more external a layer of loose connective tissue containing glands. A layer of striated muscle fibres stretches between the anterior surface of the ends of the incomplete cartilaginous rings. Between every two of these rings are strong bands of elastic and connective tissue.

THE BRONCHI.

The bronchi have essentially the same structure as found in the trachea. Lining their tubes until they are reduced to $\frac{1}{50}$ of an inch in diameter are ciliated epithelial cells. As the passages become finer and finer by their division, the cartilaginous half-rings disappear, and simple lamellæ appear in their place.

Smooth muscular fibres continue as rings round the bronchial tubes to near their points of termination. The walls grow thinner and more delicate in structure ; and when the tube is reduced to $\frac{1}{50}$ of an inch in diameter, pavement epithelium takes the place of the ciliated.   This continues until the branch is reduced so that it measures but $\frac{1}{80}$ of an inch, when it has only a basement layer of elastic fibrous tissue, lined with a thin mucous membrane covered with pavement epithelial cells. It is now known as the "ultimate bronchial tube."

It terminates in a pyramidal sac called the "pulmonary lobule," which when moderately extended is about $\frac{1}{12}$ of an inch in diameter. It represents the lung of the frog in miniature.

THE LUNGS.

First of all then we must study the simple structure of the amphibian lung. The animal is killed by breaking up the brain substance (pithing). The abdominal walls are opened in the median line with a pair of straight scissors. The under blade is then pushed up under the sternum, care being exercised that it is kept in close apposition to the inner surface lest it wound the heart. The sternum is cut through in the median line, and just beneath it lies the heart in its pericardial sac, uninjured and still beating. The mouth is opened, when a little papilla at its back part appears. This is the slit glottis of the frog, formed by folds of the mucous membrane of the mouth, in each fold of which is a cartilage of similar form, the arytenoid cartilages. Between these folds and in this slit place the end of a small blow-pipe, or what answers every purpose, a small glass pipette. If now air be forced through this tube the lungs will expand to their full capacity and rise out of and above the walls of the

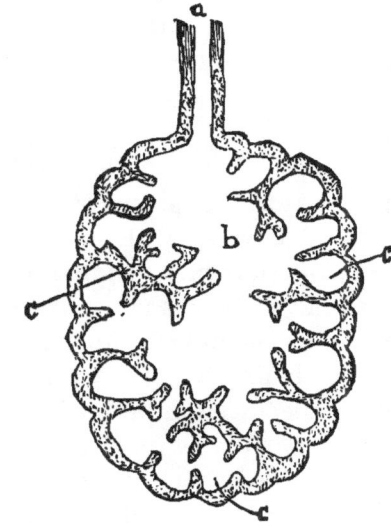

FIG. 79. Lobule of human lung. a, ultimate bronchial tube ; b, interior of lobule. c, pulmonary vesicles. x 15.

opened thorax.   An assistant now passes a thread tightly
around the base of each lung and by aid of a pair of scissors
both lungs are removed, inflated and uninjured.   If placed
near a warm stove, or in the sunlight, they soon dry and retain
their expanded form for an indefi-
nite time.   When dry they can be
opened and their internal arrangement   examined   even   with   the
unaided eye.

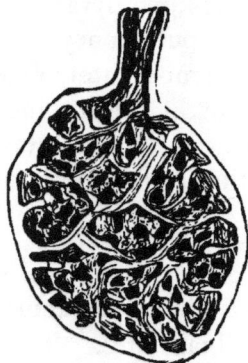

Each lung is a transparent oval
sac, pointed at its posterior extremity and covered with a layer
of   pleuro-peritoneal   membrane.
The internal surface is not smooth
as in the salamander's, and the
newt's, but is divided into a number of smaller cavities formed by

FIG. 80.   One-half frog's lung; showing interior division into air cells. Natural
size.

folds of the walls of the pulmonary sac.   These folds divide
the cavity into a number of "air cells," increasing the extent
of surface and thus giving more room for the distribution of
the branches of the pulmonary artery.   Of such a structure is
each pulmonary lobule, of the human lung.

Each pulmonary lobule is divided into compartments by
irregular partitions, on the walls of which the blood-vessels
ramify.   These compartments are known as the "pulmonary
vesicles;" they correspond to the "air cells" of the frog's lung.
Their average diameter is about $\frac{1}{100}$ of an inch.   It has been
estimated that in the lungs of man there are eighteen hundred
millions of them, exposing a surface of fourteen hundred
square feet.   The walls of the vesicles are composed of a thin
connective tissue surrounded by elastic fibres.   Unstriped
muscle is absent from the greater part of the alveolar wall, existing in small amount sometimes toward its base.   The
vesicles are lined with polyhedral cells, between which are

larger or smaller openings which lead into the lymph canalic-
ular system (Klein).   In the lungs is a rich, close net-work of

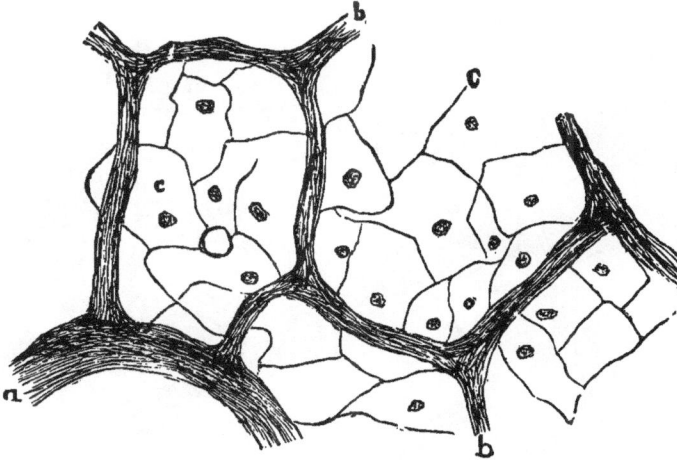

FIG. 81.   a, walls of alveolar duct.   b, walls of alveoli.   c, large flat nucleated cells
lining alveoli.   (Modified from Klein & Smith.)

capillaries.   The meshes are usually small, but vary according
to the degree of distention of the vesicle.   They are formed
from the repeated divisions of the pulmonary artery and completely encircle each individual vesicle ; giving rise in this connection to an uncommonly close net-work of tubes, which are
but slightly embedded in the alveolar walls.   The greater part
of the walls of the blood-vessels extend into the central cavity,
so that if the lung be but partly inflated, they will hang in loops
in the lumen of the alveolus.

METHODS OF EXAMINING.

In order to study the lung tissue, portions of fresh lung
may be teased or picked.   Acetic acid or alkalies added, and
examined at once.   This shows readily the elastic fibres.   The
frog's lung may be inflated and dried as described above, and

thin sections made in every direction.   The sections can be
moistened, stained, and treated with the acetic acid when quite

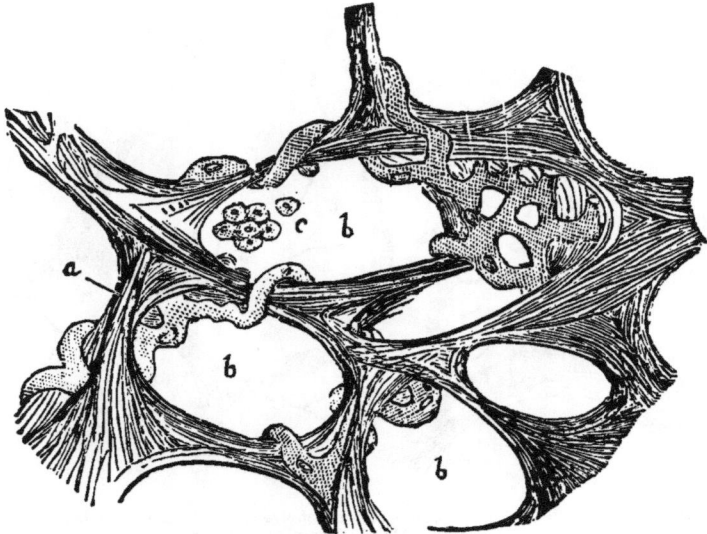

FIG. 82.   Transverse section through the pulmonary substance of a child of nine months
a, number of pulmonary cells.   b, surrounded by the elastic fibrous net-work, which bound
them in a trabecula-like manner,   and, with the thin structureless membrane, forming their
walls (a) ; d, portions of the capillary net-work with their vessels curved in a tendril-like
manner, projecting into the cavities of the pulmonary cells.   c, remains of the epithelium.
(Frey.)

satisfactory examinations can be made.   If possible however,
the blood-vessels and the air passages should be fully injected.
It is in this way only that one gets anything like an intelli-
gent view.   In the smaller animals the venæ cavæ are tied, and
the nozzle of the syringe secured in the right ventricle.   The
left ventricle is opened to allow the escape of the blood as it is
forced out by the injecting material.   At the same time the injecting mixture is flowing in the vessels the lungs are partially
expanded by means of a blow-pipe in the trachea or bronchus.
In a few moments the opening in the left ventricle is closed
to prevent the escape of any of the injecting mixture.   The
color of the lungs will decide when they are sufficiently inject-

ed.   Prussian blue has afforded all that could be desired in our
hands.   Then very small pieces of the injected lung are
placed in melted cocoa-butter and allowed to remain there for
two or three hours, when they are removed and the butter
allowed to harden.   Thin sections can be made with a razor in
every direction, with or without the aid of the microtome.
Instead of inflating the lungs with air,melted cocoa-butter may
be used.   Oil of cloves dissolves out the butter in a short time,
when alcohol can be added to get rid of the oil.   The specimens
can be stained now with carmine or logwood, cleared in oil, and
mounted in dammar.   These organs may be hardened by injecting in the trachea, or one of the bronchi, a weak solution
of chromic acid, $\frac{1}{8}$ per cent.

Enough should be injected to distend the lungs slightly,
when the trachea or bronchus should be tied and the whole
lung placed at once into some of the same solution.   The
lungs must be cut in small pieces in a few days and placed in
a fresh solution, either of the same strength or slightly increased.   In a week or ten days the pieces are transferred to
dilute alcohol and in a few hours to spirits of full strength.
Sections can be made now, stained, cleared and mounted.
The lungs of embryos should be studied especially.   Simple
hardening in alcohol, staining and clearing, will be sufficient to
enable one to recognize the structure of the whole organ.

Pigmented lungs or portions of lung are not uncommon.
A fine, black, granular pigment is observed in the walls of the
alveoli.   This may be caused by small effusions from the pulmonary capillaries ; yet it is almost invariably of extraneous
origin, being composed of minute particles of carbon inhaled
in the smoke and soot.   The lungs of those employed in coal
mines are made sometimes quite black by their inhaling particles of carbon in a finely divided state.   Some of these particles may enter the lymphatics and be carried to the lymph
glands of the bronchi, or to more distant structures.   Animals

confined in a sooty room suffer this same pigmentation of their lungs.

If it is desired to examine lung tissue after it has undergone some of the progressive inflammations, small pieces should be hardened in chromic acid or Müller's fluid and sections carefully made at the proper time.

In suspected cases of phthisis where it is very desirable to know the progress made by the disease, great aid may be procured many times by an examination of the sputa of the patient. It is now a recognized fact that phthisis has been diagnosed and is diagnosed in this way, weeks, months before other signs are manifest.

As expected ingredients in the sputa, one finds remains of food, starch granules, epithelium, air bubbles, mucous cells, pus cells, blood corpuscles, large granular cells, and perhaps, pigment cells. If now besides these there are found fragments of lung tissue, as yellow elastic fibres, it shows that there must be a disintegration of the pulmonary tissue, a condition which must denote serious trouble. If these fibres are not found it does not by any means prove that serious trouble may not exist, but their presence is very significant. Some special directions should be given to the patient whose sputa we are about to collect. First, the mouth should be carefully and thoroughly rinsed and teeth brushed after each meal. Second, the vessel in which the sputa are collected should be scrupulously clean. Third, if the patient is in the habit of using tobacco, it should be denied during the collection of the sputa, as the fibres of the leaf might mislead and cause a wrong diagnosis. If the amount of sputa is small, then all raised during the twenty-four hours should be saved. If large, that first raised in the morning should be preferred.

Any little grayish masses should be chosen and placed at once under the microscope. Acetic acid will clear up the mucous, etc., and render more distinct the yellow fibres if they

are present.   If this examination reveals nothing, the following
method should be adopted :

Make a solution of sodic hydrate, 20 grains to the ounce

FIG. 63.   Fragments of lung tissue (yellow elastic fibres).   Sputa from case of phthisis.
Mucus, pus, epithelium, granular matter, etc., were in great abundance. x 215.

of water.   Mix the sputa with an equal bulk of this solution
and boil.   Then add to this mixture 4 or 5 times its bulk of
cold water.   If possible, pour into a conical-shaped glass and
set aside.   Soon the yellow fibres, if present, will fall to the
bottom ; from here they can be drawn up with a pipette and
examined.   Several glass slides should be examined at a
single sitting, and the examination should be repeated every
few days until the presence or absence of these fibres is
satisfactorily demonstrated.

CHAPTER X.

The Salivary Glands and the Pancreas.

THE salivary glands all resemble each other in general structure, although minor differences exist. The difference in the secretions of the different glands depends upon a slight difference in their anatomical structure.

In all the glands the gland tissue is divided by connective-tissue septa into lobes and lobules. The connective-tissue frame-work surrounds the ducts and gives support to the blood-vessels, nerves, etc.

The largest (microscopical) ducts are lined with a single layer of columnar epithelial cells. In their walls are non-striated muscle cells. In the smaller ducts there is a relatively small lumen lined with columnar epithelial cells which are composed of closely arranged, thick, longitudinal rods (Kölliker.) These rods anastomose laterally to make the intracellular net-work of Klein. The small ducts branch into still smaller ones which ultimately terminate in the alveoli of the gland substance.

Just before the duct terminates it becomes narrower, sometimes branched, its lumen smaller, and its epithelial cells become a single layer of flattened cells.

The terminal alveolus is a rounded sac, convoluted and wavy in appearance. These extend in every direction. They average about the $\frac{1}{500}$ of an inch in diameter. These follicles or alveoli represent the rounded sac-like terminations of the salivary tubes.   Between these are blood-vessels, capillaries, nerves, lymphatics, and connective tissue.

In the parotid gland—the only true salivary gland—and in most of the lobules of the submaxillary, the lumen of the alveoli is small and lined with a single layer of glandular epithelial cells, which are columnar in shape but very short. During secretion the lumen is smaller and the lining cells are broader.   In a few of the lobules of the submaxillary and in

FIG. 84.   Two salivary tubes from the lobule of a muciparous gland, entering the main duct. a, duct of the lobule. b, salivary tube. c, follicles, on one side, as they appear *in situ*. d, follicles separated from each other, showing the windings and offshoots of the salivary tube.
(Kölliker.)

nearly, if not all of those of the sublingual the alveoli are larger and lined with two kinds of cells.   1, The mucous cells are like the goblet cells already described.   These contain a transparent substance called mucigen (Heidenhain), which during the secretion of the gland is transformed into mucin.   2. The crescents of Gianuzzi.   These are composed of nucleated, polyhedral, granular cells, smaller than the mucous cells and situated in the periphery of the follicle.   They are arranged very close together after the fashion of a crescent. They represent young mucous cells.   After exhaustion of a gland, after the mucous cells have given off their mucin, they disappear, and now only a granular substance remains in the body of the cell.   The alveoli are surrounded by a fine capillary net-work of blood-vessels and lymphatics.

A net-work of fine nerves is seen in connection with the intra-lobular connective tissue.   In this net-work are numerous ganglia composed generally of unipolar ganglion cells.   The fine nerve branches terminate in the epithelial cells lining the salivary tubes, each cell representing the terminal organ of each nerve fibre.

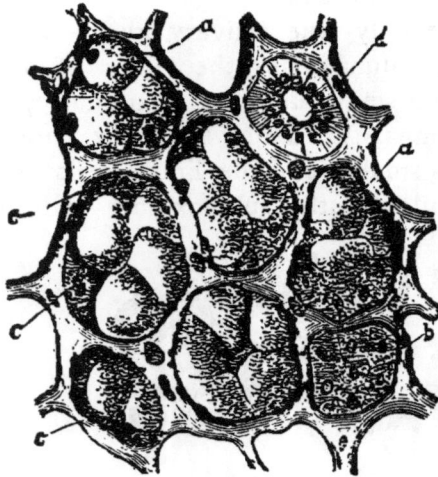

FIG. 85. The submaxillary gland of the dog. a, mucous cells. b, protoplasma cells. c, crescent. d, transverse section of an excretory duct, with the peculiar cylindrical epithelium. (After Heidenhain.)

METHODS OF EXAMINING.

Small pieces of the gland are placed in a saturated solution of picric acid for 48 hours. Sections can then be made with a freezing microtome.

Heidenhain employs absolute alcohol and stains with carmine.   Small pieces can be macerated in, at first, a weak solution of chromic acid $\frac{1}{32}$ grain to the ounce of water, and gradually increased to ½ grain to the ounce.   The ducts are best injected with the Prussian blue.

Osmic acid affords a useful reagent, especially for the study of the epithelial cells.   A .5 per cent. solution should be employed, and small pieces of the tissue allowed to remain in it for 48 hours.   If this reagent is used for hardening, a weaker solution (.25 per cent.) should be used, and the tissue allowed to remain in it for 2 or 3 days, when sections can be made with the freezing microtome.

THE PANCREAS.

The pancreas was formerly known as the "abdominal

salivary gland." It is divided into lobes and lobules and in its intimate structure resembles the salivary glands. The tissue is some softer than that belonging to these glands ; this is simply because the lobes are not so compactly arranged. The cells in the alveoli resemble those of the parotid.

CHAPTER XI.

The Pharynx. Œsophagus. Stomach and Intestine.

THE PHARYNX.

THE walls of the pharynx are composed of transversely-striated muscles with fibrous connective tissue. They are lined by a mucous membrane which is covered with a layer of epithelial cells, columnar and ciliated in the upper third, but squamous and destitute of cilia below. The mucous membrane is rich in mucous glands which, in the upper part, collect in groups with a common excretory duct.

THE ŒSOPHAGUS.

At about the commencement of the œsophagus (cricoid cartilage) the striated muscle walls of the pharynx are replaced by the smooth muscle tissue until very soon there are two muscular layers, an external longitudinal, and an internal circular. This arrangement of muscle extends throughout the remainder of the alimentary canal. Beneath the internal circular layer is a coat of submucous areolar tissue, in which are the glands opening on the surface of the mucous membrane. The mucous membrane is quite thick and is arranged in longitudinal folds when the œsophagus is not distended. It projects into the epithelial layer as small cylindrical papillæ, in the centre of which may be a lymphatic, terminating in a loop or in a single blind vessel. A thick layer of flattened epithelium covers the membrane and extends to the cardiac orifice of the stomach. Between the mucous and submucous coats is a longitudinal layer of smooth muscle fibres, which forms a continuous coat at the lower part of this tube. This is the muscularis mucosæ.

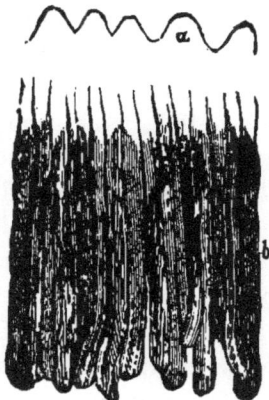

FIG. 86. Vertical section of the human gastric mucous membrane ; a, surface papillæ. b, glands. (Frey.)

THE STOMACH.

The stomach is surrounded by a serous coat derived from the peritoneum, beneath which is a smooth muscular layer, continued directly from the œsophagus and arranged the same as described for that tube. Still deeper than this is another layer of muscular tissue arranged in an oblique manner. Between the muscular and mucous coats is a quantity of areolar tissue in which are found the blood-vessels, fat cells, etc. The mucous membrane is at once very complex and very interesting. It is loosely attached to the muscular coat and is easily movable over it; so that when the stomach is empty it is thrown into numerous folds, or rugæ, which have a general longitudinal direction, and are most marked along the great curvature. They are entirely obliterated when the organ is distended. Lining the mucous membrane over the whole surface of the stomach is a layer of columnar

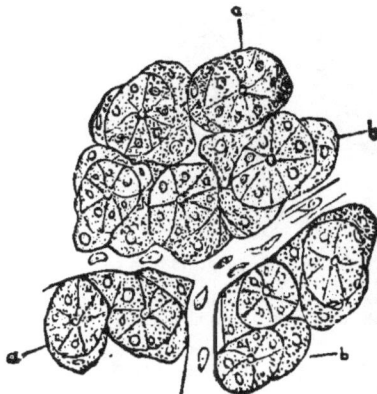

FIG. 87. Horizontal section through fundus of stomach of dog. The peptic glands are cut transversely. a, cells lining lumen of gland. b, parietal cells. (after Klein and Smith.) x 400

cells $\frac{1}{800}$ to $\frac{1}{100}$ of an inch high, and $\frac{1}{4000}$ to $\frac{1}{5000}$ of an inch broad. Among these cylindrical cells are a few that are goblet-shaped; especially is this true during digestion when many of these cells are seen. The epithelium extends into the ducts of the glands which are placed perpendicularly to the surface. Beneath the layer of epithelium is a basement membrane of flat nucleated endothelial cells. In the mucous membrane are found connective-tissue cells, endothelial cells, lymph cells and the glands and their tubes. If the surface of this membrane be examined with the aid of a common lens small depressions are seen about the $\frac{1}{150}$ of an inch across. On these alveoli are the small round openings of the ducts of the tubular glands. Two kinds of glands are imbedded in the mucous membrane; those in the pyloric portion—pyloric glands, and those in the other portions, —peptic glands.

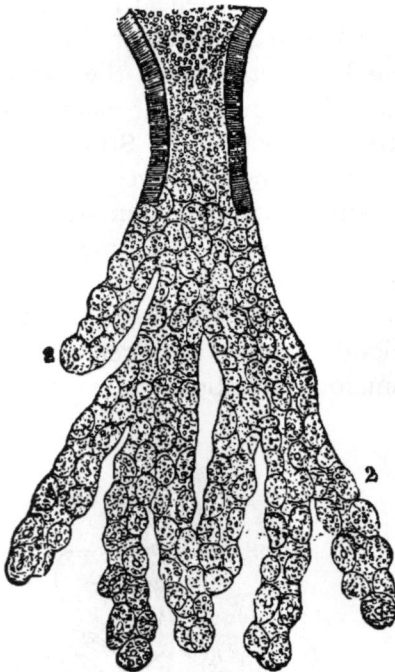

FIG. 88. Peptic gland from cardiac portion of human stomach. 1, Excretory tube, leading to the surface. 2, Tubular follicles, containing spheroidal cells. [Kölliker.]

The peptic glands consist of three or four gland tubes, closed at their deep extremity, which empty into one common duct. A gland is divided into three parts, its duct, neck and body. The duct extends from $\frac{1}{5}$ to $\frac{1}{3}$ the whole length of the tube; the neck from $\frac{1}{4}$ to $\frac{1}{3}$ of the same distance. Both the duct and neck are lined with the same cells that cover the free surface of the mucous membrane, with this difference, that in the neck the cells are shorter and the nuclei smaller. Outside of these cells, but inside the walls of the tube, are other nucleated cells of a granular appearance. They do not form a continuous layer, being found only here and there. For a long time they were described as "peptic cells," but Heidenhain showed their position in the tube, and gave them the name of "parietal cells."

The body of the gland is lined with a layer of cells directly continuous with those of the neck, only here the cells are more columnar. They have a spherical or oval nucleus which takes staining very readily. But few parietal cells are seen in the body of the gland, and they decrease very rapidly towards the fundus (Rollet.) During digestion the cells in the body of the glands become thicker, more opaque and granular, giving the whole gland a broader appearance (Heidenhain.)

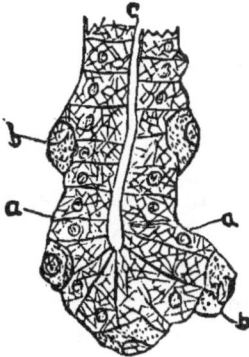

FIG. 89. Fundus of a gland tube. The chief cells a, have a distinct reticulum. b, the nucleated parietal cells. c, lumen of tube. [After Klein.] x 450.

As these glands approach the pyloric portion of the stomach, the ducts become longer and the remaining portions shorter. It has been noticed in some animals that when the peptic and pyloric glands meet, they not only intermix, but also that the peptic glands become transformed into the pyloric.

The pyloric glands have a much longer duct, with a very short neck. The body is composed of two or more tubes. The lumen of the gland is many times larger than in the other glands. No parietal cells are found in these glands. By some they are regarded as simple peptic glands, while this is denied by others; they are certainly not mucous glands as was formerly supposed.

These glands continue to the pyloric orifice, and then pass through into the duodenum as Brunner's glands, which are

identical with them in structure. A layer of muscle fibres
separates the mucous from the submucous tissue (musculrisa
mucosæ.)

The fine terminal arterial branches enter the mucous membrane and form a plexus upon the walls of the gland; branches
of this plexus surround the mouths of the glands and borders
of the alveoli; (see figure 72.) The lymphatics arise in the
mucous membrane by a net-work of vessels, situated between
the tubules. They are not as superficial as the blood-vessels,
and they commence as loops, or in dilated extremities. Beneath the mucous membrane they empty into a fine plexus,
then pierce the muscularis mucosæ to form another coarser
plexus in the submucous coat; the branches then penetrate the
muscular layers and follow the direction of the blood-vessels
until they reach some of the lymphatic glands found on the
surface of the stomach.

Fine nerve gangliated plexuses are seen between the muscular layers and in the submucous coat.

THE INTESTINE.

The walls of the intestine, both small and large, are the
same as those of the œsophagus and stomach, viz.: 1, an epithelial layer resting on 2, a mucous membrane beneath which
is 3, the muscularis mucosæ which in turn is surrounded by
4, a submucous tissue, and external to this are 5, the circular,
and 6, the longitudinal layers of muscle cells. A serous
covering, from the peritoneum, surrounds the whole.

THE SMALL INTESTINE.

The mucous membrane of the small intestine is thrown
into a number of folds which cannot be obliterated by the distention of the canal. These are the valvulæ conniventes.
They extend one-half or two-thirds of the distance around the
canal, are about ⅓ of an inch wide and are placed in close

succession. Small processes of the mucous membrane cover these folds completely, and fill the spaces between them. These prolongations are the villi; they are from $\frac{1}{30}$ to $\frac{1}{40}$ of an inch long, and are most numerous in the duodenum and jejunum. Krause estimates the whole number in the intestine to be at least four millions. Columnar epithelial cells cover the villi and the mucous membrane of the small intestine throughout its entire length. At the base of each villus is a small arterial branch which

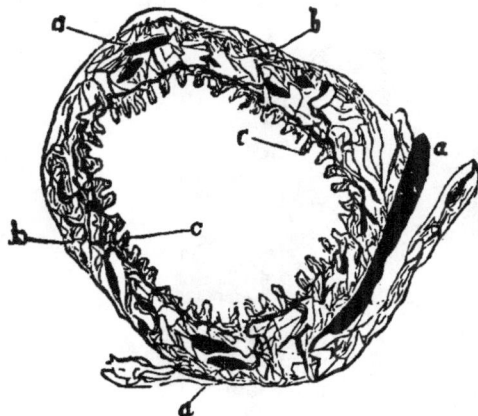

FIG. 90. Transverse section of the ileum of an infant, injected. a, blood-vessels filled with the injecting material, b, muscle walls. c, villi.

runs in the centre toward the surface. It soon gives off a number of branches to make a most beautiful capillary network which terminates in a vein, and this in turn penetrates the mucous coat to pass into the submucous layer (see figure 73.) The small lymphatic or lacteal, occupies the centre of the villus, usually as a single tube, sometimes double, with walls of a single layer of endothelial cells or plates. It usually commences as a blind tube, with perhaps a dilated extremity and surrounding it are a few muscle cells arranged in a longitudinal manner. These cells are prolongations from the muscularis mucosæ and when stimulated they cause a decided retraction of the villus.

The matrix of each villus is composed of a delicate reticulum in which are seen a few nucleated fat cells. This reticulum forms the interstitial substance of the basement membrane, between the covering of epithelial cells and the mucous membrane; from this, prolongations extend between the epithelial cells on the surfaces of the villi; it forms the interstitial substance between the endothelial plates composing the wall of the lymph capillaries, and also of the blood capillaries.

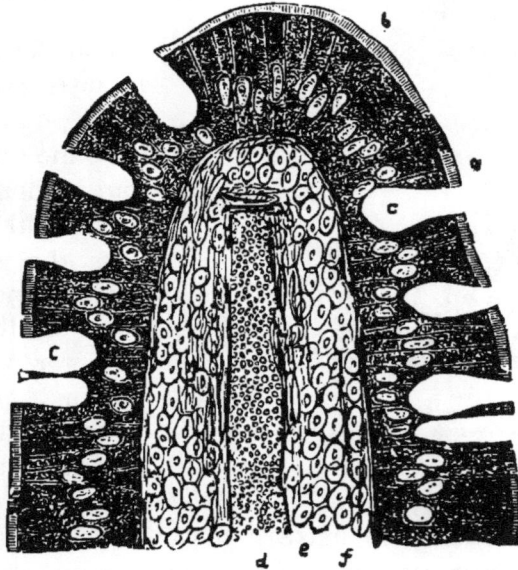

FIG. 91. Vertical section of a villus of the small intestine of a cat, hardened in chromic acid. a, streaked basal border of epithelium. b, cylindrical epithelium. c, goblet cells. d, central lymph vessel. e, smooth muscular fibres which lie nearest to the lymph vessels. f, adenoid stroma of the villus in which lymph corpuscles lie. (Klein.)

A very interesting question presents itself here as to the manner in which chyle passes from the interior of the intestine into the lacteals and blood-vessels. It is quite generally taught that the minute chyle globules pass directly into the body of the cells covering the villi, and thence pass into the capillaries or lymph ducts. Thus Kölliker and others have described and illustrated these particles of oil as passing through the cells. The researches of Dr. Watney, under the direction of E. Klein, have thrown much needed light upon this subject. According to these authorities the epithelial cells are engaged here in secreting mucus and not in absorption. Absorption takes place through the interstitial substance of this delicate reticulum. The chyle globules then pursue the following course: They enter the interstitial substance between the epithelial cells and pass into the directly continuous basement membrane; from here they pass into the matrix reticulum of the villus, and finally into the interstitial substance between the endothelial plates into the blood or lymph channels.

Thus it is seen that there is a continuous lymph-canalicular system from the free border of the villus to the central vessels. This affords a very satisfactory explanation of the absorption from this canal, for, owing to the centripetal direction of the flow in the lymphatic vessels, there must be a strong tendency for matter outside the vessels to pass into their interior. At the base of the villi are the crypts of Lieberkühn. These are minute tubes placed perpendicularly to the surface, and consist of tubes lined with columnar epithelium. They are from $\frac{1}{250}$ to $\frac{1}{350}$ of an inch in length and $\frac{1}{600}$ of an inch broad. In

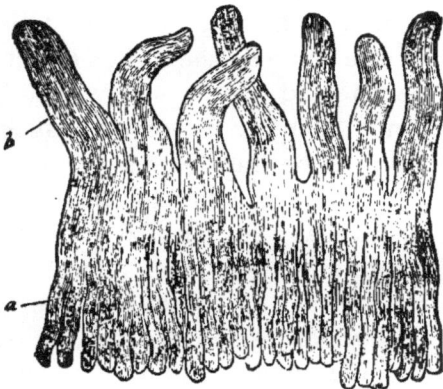

FIG. 92. Lieberkühnian glands (a) of the cat, with the intestinal villi (b) situated over them. (Külliker.)

the submucous tissue of the duodenum are the glands of Brunner, identical with the pyloric glands.

Between the longitudinal and circular layers of muscle is a plexus of band-like nerve branches, among which are groups of glanglion cells, called the "plexus mesentericus of Auerbach." Branches pass from this plexus into the submucous tissue to give rise to the "plexus of Meissner." Here are unipolar, bipolar, and multipolar ganglion cells.

Branches are distributed from both these plexuses in **every** direction.

THE LARGE INTESTINE.

The walls of the large intestine resemble in all respects those of the small. The mucous membrane, however, is destitute of villi, but more freely supplied with the crypts of Lieberkühn. Here these glands are placed more closely together, are longer and more numerous.

METHODS OF EXAMINING.

There are many methods employed to demonstrate the structure of these organs. Only stomachs from recently killed animals should be chosen, for if only a short time has elapsed since the death of the animal, the softer structures of the inner coats will be affected.

A piece of stomach an inch square, or several such pieces are placed in Müller's fluid, to be transferred in a week or ten days to alcohol. Vertical sections show all the coats to good advantage. Small pieces can be hardened in alcohol alone, and stained with carmine. Sections should be made perpendicular to the free surface of the mucous membrane, in order to obtain good views of the glands. For this purpose, small pieces from different parts of the mucous membrane are placed immediately in absolute alcohol. When hardened, vertical sections are made as thin as possible, and stained in carmine or hæmatoxylin. The sections are well preserved in glycerine or dammar. Horizontal sections show the different parts of the glands cut across. They are to be stained and mounted as the others. Small pieces of the mucous membrane are placed in a ¼ per cent. solution of osmic acid for 24 hours and sections made with the freezing microtome. This process demonstrates the cells in the glands in a very satisfactory manner.

To study the blood-vessels of the organ, the entire animal, if small, should be injected. Beautiful preparations are made from the stomach of a cat or rabbit, injected with Prussian blue or carmine.

Pieces of the small intestine are treated like those of the stomach, only in the injected specimens sections should be made with the aid of a freezing microtome in order that the villi may not be injured. Such specimens should be stained but slightly. To examine the lacteals, the animal should be given a meal composed largely of fatty matters, and then killed in 3 or 4 hours. Small pieces of the mucous membrane are placed in a 1 per cent. solution of osmic acid and allowed to remain therein nearly 48 hours. This stains all the fatty particles black, and thus shows the outlines of the lacteal in the centre of the villus. Chromic acid, chromate of potash, teasing, penciling, and tingeing should not be forgotten

CHAPTER XII.

The Liver.

THE liver is very early to show itself, for it has attained comparatively an enormous size by the end of the first month of uterine life. Buds or projections appear on either side of the intestine to form the two principal lobes of the liver. It soon occupies nearly the whole of the abdominal cavity and in proportion to the body weight, bears the following relation according to Berlach:

At the end of the first month as 1 to 3. At full term as 1 to 18. In the adult as 1 to 36. It is very soft in structure during the first months, but as its development progresses its tissue becomes more firm. The liver is very peculiar in its structure, and is unlike any of the other glands of the body.

By the unaided eye one is just able to discern on the natural external surface of the human liver an innumerable number of pentagonal or hexagonal islets, known as the hepatic lobules. They are about $\frac{1}{12}$ of an inch in diameter and their number corresponds with the number of central veins. In the liver of the pig these lobules are easily distinguished by the naked eye, for a considerable amount of connective tissue completely separates them from one another. In the human liver this connective tissue is not nearly so well developed, and in many parts of the organ the surface of one lobule is in direct contact with the adjacent ones.

Each lobule is composed essentially of two substances :

144

the liver cells and the capillaries. The capillaries present an uncommonly complicated net-work. Commencing with the portal vein, it is seen to enter the transverse fissure together

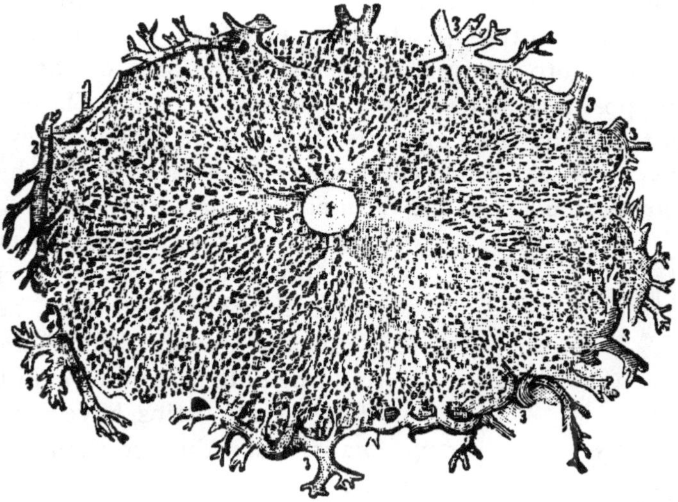

FIG. 93. Transverse section of a single hepatic lobule. (Sappey.) 1. intralobular vein, cut across. 2, 2, 2, 2, afferent branches of the intralobular vein. 3, 3, 3, 3, 3, 3, 3, 3, branches of the portal vein, with its capillary branches, forming the lobular plexus, extending to the intralobular vein.

with the hepatic artery, hepatic duct, nerves and lymphatics, surrounded and bound together by a fibrous connective tissue, the capsule of Glisson. From the portal vein are given off small branches which pass between adjacent lobules. As these terminal branches extend only between the lobules and never enter within them, they are called the intermediate veins, interlobular veins. From these veins branches are derived, which rapidly divide into a close capillary net-work, and approach the centre of the lobule in a radial manner. The net-work is formed by horizontal or transverse branches connecting the principal capillaries as they pass from the periphery to the centre of the lobule. This forms a "lobular plexus" of capillaries, with each vessel about the $\frac{1}{2000}$ of an inch in

diameter.   As they near the centre of the lobule, they unite
together to form veins of considerable size, which empty into
a single central vessel placed in the long axis of the lobule.
This vessel, from its central position, is called the intralobular
vein.   It is from the $\frac{1}{1000}$ to the $\frac{1}{400}$ of an inch in diameter
and empties into a larger vessel just at the base of the lobule,
from its position, named the sublobular vein.   These veins
collect the blood from all parts of the liver, and convey it to
larger vessels, which become larger still, until the three hepatic
veins are formed.   At the termination, then, of the portal vein,
are the interlobular branches, and
the intralobular vessels mark the
commencement of the hepatic
veins, the lobular plexus being the
intermediate system of capillaries.
The hepatic veins have very thin
walls, much thinner than the portal, and they are not surrounded
by any Glisson's capsule, but are
quite firmly united to the hepatic
tissue.   The hepatic artery enters
the sheath at the transverse
fissure, and immediately gives off

FIG. 94.   a, small hepatic vein. b, sublobular veins.   c, lobules. (Kiernan.)

branches to the walls of the portal vein, and a very rich plexus
to the walls of the hepatic duct, so that when this artery is
thoroughly injected it almost covers, with its capillaries, the
walls of the duct.   It supplies also the capsule of Glisson, the
branches known as capsular branches.
     Branches of this artery are interlobular, and accompany
the branches of the portal vein.   According to Beale, these
branches lead into special veins, which accompany the arteries
in couples, and join the interlobular branches of the portal.
According to some observers, the capillaries and veins from
the hepatic artery join the capillaries of the lobules by an

anastomosis of the blood-capillaries of the bile ducts with the capillaries of the lobules.

All the space not occupied by the capillary net-work is

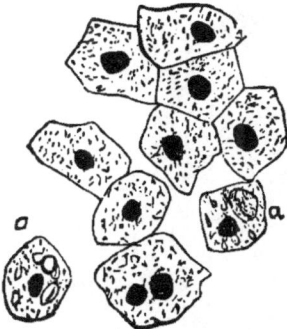

filled with the glandular cells of the organ. These are circular when seen single, but angular, pentagonal, or hexagonal when viewed in thin sections of the liver. They are about the $\frac{1}{1000}$ of an inch in diameter, and are provided with a well marked nucleus. Here again is the inter-cellular and intranuclear fibrillar net-work of Klein and others. Some of the cells have two nuclei, and minute fat globules, one or more to

FIG. 95. Liver cells. a, contain-ng fat. x 400.

to each cell are very generally seen. The fatty embedments are present in the liver cells of adults whose diet is rich, and they occur also in the young infant, and in the fattened lower animals. The liver cells may become in this way crowded with fat, which soon dissappears when the manner of living is changed. Small particles of pigmentary matter are sometimes abundant in the cells, giving them a peculiar brownish appear-ance. Each cell is in direct contact by some of its surface with a blood-vessel.

It is with the greatest difficulty that the bile capil-laries can be investigated. A fine system of bile ducts is recognized running with the interlobular branches of the por-tal vein. Into these fine branches empties a delicate net-work of the finest biliary capillaries, their diameter varying from the $\frac{1}{26000}$ to the $\frac{1}{12000}$ of an inch. These capillaries run between the liver cells, and thus enclose polygonal spaces of the shape and diameter of a single cell. In this way every liver cell, from the periphery to the centre of the lobule, is in contact, by one part or another of its surface, with these fine biliary pass-

ages.   These capillaries always pursue a course with reference
to their keeping, as far as possible, from the blood capillaries,
for there always remains part of a cell between the bile capil-

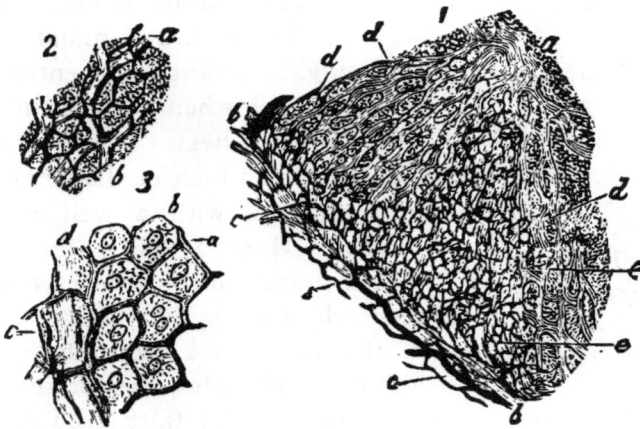

FIG. 96.   Biliary capillary of the rabbit's liver.   1. A part of the lobule; a, vena hepatica.
b, branch of the portal vein. c, biliary ducts. d, capilliaries.   2. The biliary capillaries (b)
in their relation to the capillary blood-vessels (a).   3. The relation of the biliary capillaries
to the hepatic cells. a, capillaries.   b, hepatic cells.   c, biliary ducts. d, capillary blood-
vessels.   (after Frey.)

lary and blood capillary.   These biliary canals pass between
the boundary surfaces of two neighboring cells, and not simply
along their borders; in this way the surface of a cell may be
divided into two equal or unequal parts.   Whenever a bile duct
is found along the border of a cell, it will be noticed that no
blood capillary comes in contact with that part of the cell in
any way.   It is impossible, then, to find a fine bile passage and
a blood capillary without intervening cell substance, and more
than this, if a few cells exist in the liver not touched by blood-
vessels, they are sure to come in contact with a fine gall duct.

Authorities are about equally divided on the question of
the bile capillaries having a membrana propria.   Some hold
that the liver cells themselves form the walls of the ducts, while
others describe and illustrate these finest ducts as possessing a

very tenacious, but delicate membrane. The latter view, we believe, is the correct one.

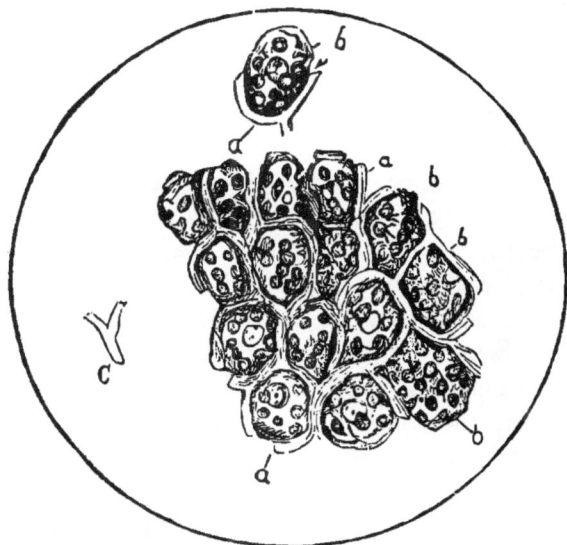

FIG. 97. From investigations of H. R. Stiles on the Texas cattle disease. a, bile capillaries. b, liver cells. c, membrana propria of the bile capillary.

METHODS OF EXAMINING.

The usual methods are followed in studying this organ, viz.: Injecting, hardening and staining. To examine the unin-

FIG. 98. From Quarterly Jour. Med. Science (Hayes) 1879, July, by W. G. Davis. a, liver cells. b, bile capillary, having a membrane.

jected liver, small pieces are placed at once in Müller's fluid and allowed to remain there for two weeks, at least. A large amount of the fluid should be used. The pieces are then placed in dilute alcohol for a day or two, and then in cohol of full strength for two or three days longer. Sections can be made now without trouble, and stained with hæmatoxylin. Sections should be made in at least two directions:

First, parallel to the surface, in order to cut the intralobular veins across, and show the arrangement of the lobular plexus of capillaries and their relation to the liver cells. Second, vertical to this surface, showing the intralobular veins along their length, and their terminations in the sublobular vein. These vessels are seen to much better advantage in injected specimens. The liver of the rabbit or cat is especially suitable for injection. The animal should be killed by bleeding. A branch of the artery, or portal or hepatic vein, or branch of a duct may be injected, or several of these may be injected at the same time with different colored mixtures. As in the case of many of the vessels of the different organs, so here especially will it be wise to make a slit through the walls of the vessel in a longitudinal direction, and insert the pipe into this rather than into the opening of the cut end of a vessel. In this way a pipe can be inserted into a vessel scarcely larger than the pipe itself. To inject the portal vein thoroughly, first a solution of salt should be injected in order to wash out the blood.

To inject the biliary passages, the pipe of the injecting apparatus is secured in the common bile duct, and slight pressure employed. The Prussian blue mixture is satisfactory here, and as soon as a few of the lobules are seen colored by the blue, the duct should be tied, and that part of the organ placed in Müller's fluid, afterwards in spirits, and finally stained slightly with hæmatoxylin, cleared in oil of cloves, and mounted in dammar.

CHAPTER XIII.

Kidney.

THE first trace of a urinary apparatus is found at an early period of embryonic life, and consists of two organs known as the Wolffian bodies; these are fully developed by the end of the first month and hardly to be detected after the second.

FIG. 99. Diagram of the formation of the uro-genital organs. 1. a, intestinal canal with protuberance b. 2. The protuberance is very much developed. a, allantois. b, the urachus. c, the bladder. d, the genito-urinary sinus, with three protuberances. 1. duct of Müller. 2. Wolffian body. 3, the kidney. (after Küss.)

At this early period the alimentary canal is a blind tube from which is given off a diverticulum. From this canal three diverticula arise. From the anterior one will be developed the female generative organs —uterus, fallopian tubes, and the analogous organs in the male— prostatic vesicle and appendage to epididymis. From the posterior one will be developed the kidneys of the adult. Between these two will arise the Wolffian bodies.

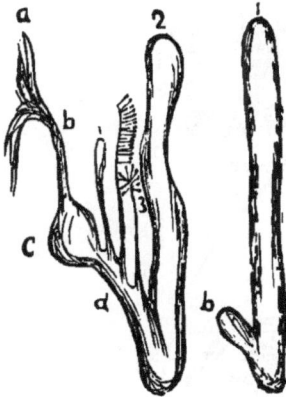

When fully developed, they will correspond in structure to the true kidney ; the tubules however will be about four times as large. The physiology of these bodies is not understood, but it is altogether probable that they answer the same purpose in the embryo that the kidneys do in the adult, viz.: They throw off from the body some principle that would be injurious to it if retained.

151

At this early history in the life of the embryo, the Wolffian bodies, liver, and intestine fill the abdominal cavity, and they are the only organs of any size present. After the first month these bodies commence to diminish in size while the organs behind them as rapidly increase in size. The result of this is, that at the end of the second month the order has been reversed, and now the true kidneys rise from behind the temporary or false ones, and by their rapid growth leave the atrophied Wolffian bodies at their lower parts.

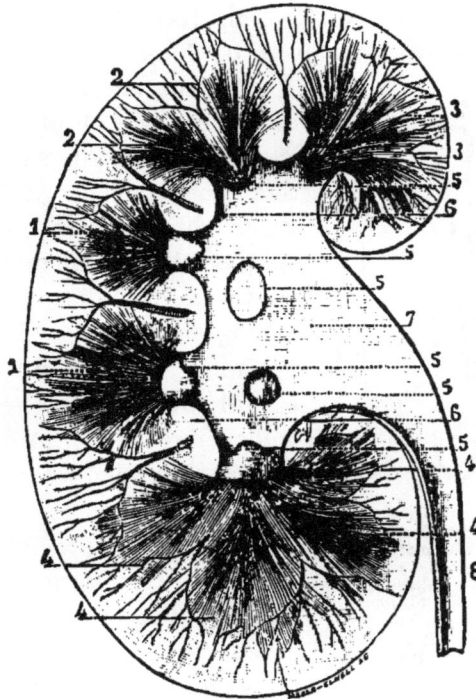

FIG. 100. Vertical section of the kidney. (Sappey.) 1, 1, 2, 2, 3, 3, 3, 4, 4, 4, 4, pyramids of Malpighi. 5, 5, 5, 5, 5, 5, apices of the pyramids surrounded by the calices. 6, 6, columns of Bertin. 7, pelvis of the kidney. 8, upper extremity of the ureter.

The kidney of the adult is likened to a bean in shape; the concavity of the bean representing the hilum of the kidney.

At this place the ureter receives the urine, and the blood-vessels find their entrance and exit.

A vertical section through this organ, from its convex to its concave borders reveals from twelve to twenty eminences projecting into the pelvis ; these are the papillæ. On each papilla there are from twelve to twenty openings which represent the terminations of that number of collecting tubes.

The unaided eye readily discerns a difference in structure between the part of the organ lying towards its concavity and the part towards its convexity. The part toward the papillæ has a fibrous parallel arrangement, while the other part is darker, homogeneous, or granular in appearance. The former is the medullary and the latter the cortical substance of this organ.

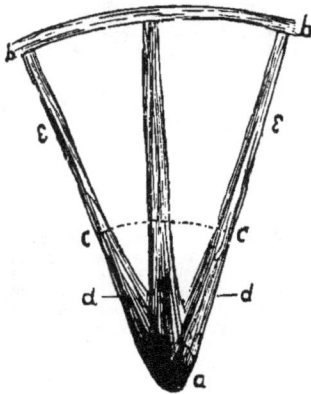

FIG. 101. Diagram illustrating the medullary and cortical pyramids. a, papilla. b, external cortical layer. c, boundary line between the medullary, and cortical substances. d, straight uriniferous tubes. e, medullary rays.

The medullary substance can be divided into pyramids; each pyramid having a papilla for its apex and an imaginary line between the medullary and cortical portions for its base. If one of the twelve or twenty tubes forming each papilla be examined, it will be seen to divide into two or more branches, each of these branches dividing again until they are reduced in size to about the $\frac{1}{500}$ of an inch. This process of division is generally complete by the time the tubes are one-fourth of an inch from the papilla; the tubes then proceed in a straight course to the extreme border of the cortical substance, unchanged in their diameter. These canals are collected into bundles, and as they are removed from the papilla, the bundles become separated from each other; as a result of this, there is a long, narrow pyramid between them and the cortical pyramid.   The base of this pyramid is formed by the external cortical layer, the sides are bounded by the straight canals and the apex extends past the dividing line into the medullary pyramid.

The straight canals that are

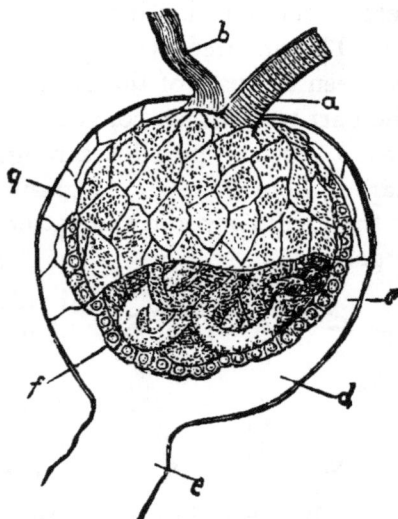

FIG. 103.   Glomerulus of the rabbit (diagramatic.)   a, vas afferens.   b, vas efferens.   c, glomerulus.   d, under portion of the capsule (without epithelium).   e, neck,   f, epithelium of the glomerulus; and g, that of the inner surface of the capsule after treatment with silver.   (From Frey.)

known in the medullary portion as the *tubuli uriniferi recti* are known in the cortical portion as the medullary rays— *medullaries radii*.

A medullary pyramid is composed, then, of straight uriniferous tubes,   straight blood-vessels, nerves, a few lymphatics, and the apices of the cortical pyramids.

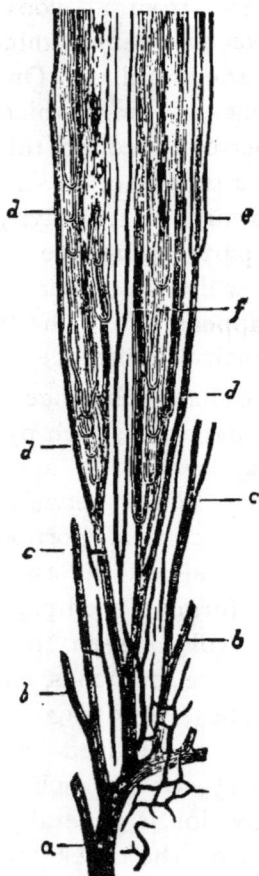

FIG. 102.   Vertical section through the medullary pyramids of the pig's kidney (semidiagramatic). a, trunk of a uriniferous canal, opening at the apex of the pyramid.   b and c, its system of branches.   d, loop-shaped uriniferous canals.   e, vascular loop, and f, ramification of the vasa recta.   (Frey.)

In the centre of each cortical pyramid is a branch of the renal artery from which arise many branches; these soon break up into a system of capillaries,—which do not anastomose,—to make the Malpighian bodies or glomeruli.

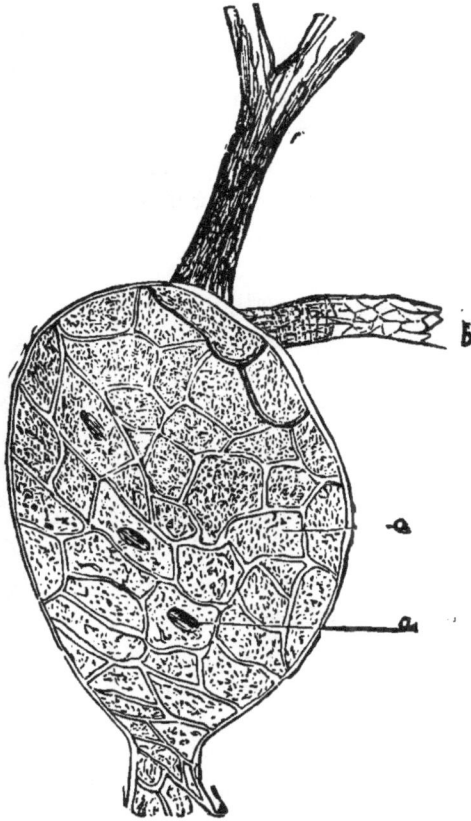

Surrounding this system of capillaries is a thin capsule, composed of two layers of cells. The cells of the inner layer are large and flat, while those of the outer layer are smaller and not so flat. This capsule of the glomerulus, Bowman's capsule, is the dilated commencement of one of the uriniferous canals. It becomes constricted at a point opposite the entrance of the arterial branch, to make the neck of the capsule. It then dilates into a broader tube, becoming convoluted for a short distance when it is arranged in a more or less spiral course; its direction is then towards the medulla, when it shortly assumes a straight direction to the apex of the cortical pyramid, forming the descending side of the loop of Henle. It now turns sharply on itself—the loop of Henle—and pursues a straight direct

FIG, 104.   Capsule of the glomerulus, rabbit's kidney. Silver treated and tinged with carmine. a, endothelial cells. á, nucleated. b, vas afferens.   vas efferens. (after Ludwig.)

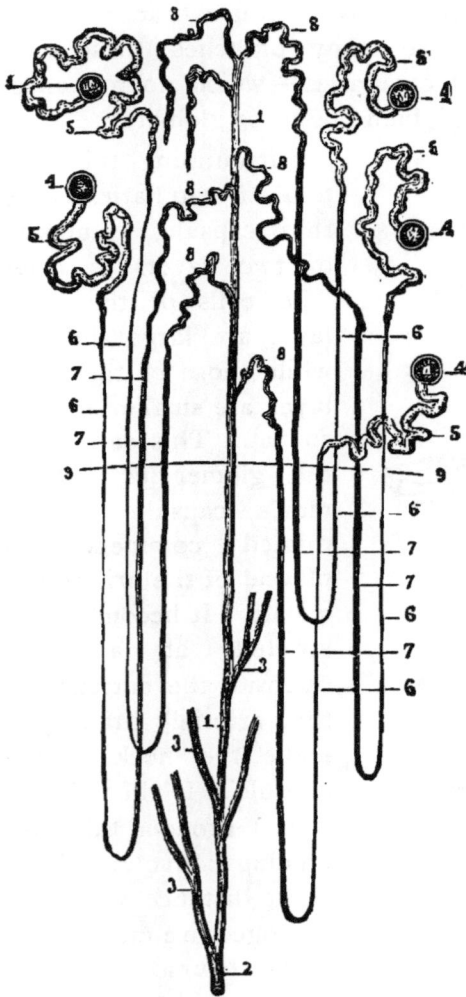

FIG. 105. Diagrammatic view of the Malpighian bodies and tubes of the kidney. (Sappey.)
1, 1, 2, straight tube of Bellini. 3, 3, 3, other straight tubes opening into the tube 1, 1. 4, 4, 4, 4, 4, Malpighian bodies. 5, 5, 5, 5, 5, convoluted tubes. 6, 6, 6, 6, 6, descending portions of the looped tubes of Henle. 7, 7, 7, 7, 7, ascending, larger portions of the tubes of Henle. 8, 8, 8, 8, 8, 8, communicating tubes. 9, 9, dotted line showing the limits of the cortical and of the pyramidal substance.

course toward the external cortical layer, forming the ascending side of the loop of Henle. It is then again arranged in a more or less spiral course, becoming later very irregular and possessed of a small lumen. It now pursues a convoluted course and becomes united with a medullary ray by an intermediate portion—the connecting canal—the intercalary portion. The medullary ray soon becomes a straight uriniferous tube, and opens on the free surface of a papilla; the union of these two makes one collecting tube.

From the dilated commencement of the uriniferous tube at the Malpighian bodies to their termination in the medullary rays, they change their diameter no less than seven times. 1st, at the constricted portion, the neck. 2nd, enlarged, in the convoluted por-

tion.   3rd, constricted in the descending side of the loop.
4th, enlarged in the ascending side of the loop.   5th, constricted at the commencement of the convoluted portion. 6th,
enlarged during that portion.   7th, constricted at the termination in the medullary rays.

The epithelial cells lining the tubes vary in different
parts. With the exceptions of the descending side of the
loop of Henle, the loop itself, and the collecting tube, all
parts of the uriniferous tubules are lined with epithelial cells,
composed of a substance which exhibits "rods or fibrils arranged vertically to the long axis of the tube." (Heidenhain.)

In the convoluted portions the cells are nucleated,
polyhedral, of unequal size and vertically striated.   The
spiral portions are lined with irregular, striated cells; some
are thin columnar cells, some having concave and others
convex sides.   These constitute the "fungoid cells" of
Schachowa.

The descending side of the loop of Henle, and the loop
itself are lined with flat, thin, nucleated cells, and Klein com-
pares this structure to that of the capillary blood-vessels,
only slight differences existing.

Klein notices a deposit of brownish pigment granules in
the epithelial cells of the spiral portion of the ascending side
of the loop of Henle.

The epithelial cells lining the collecting tubes are nucleated, polyhedral, spindle-shaped, flat or angular.

With the exceptions of the descending side of the loop of
Henle, and the loop itself, there is a membrane covering the
inner surface of the epithelial cells of all the urinary tubules.
This membrane is next to the lumen of the tube and is called
the "centro-tubular" membrane. (Klein and Smith).

A study of the physiology of this organ teaches that the
water of the urine is taken from the blood at the glomeruli,
that it passes from the capillary system into the dilated com-

mencement of the urinifereous canal, and at last reaches the
free opening on the surface of the papillæ.   Klein has dis-
proved the ideas of Heidenhain concerning the excretion of

pigment, and has
demonstrated that
carmine in ammonia
injected into the
circulating blood of
the cat, will be de-
posited between,
and not in, the epi-
thelial or endothelial
cells themselves.

The solid consti-
tuents of the urine,
then are excreted
along the course of
the uriniferous
tubes.

A scanty frame-
work of connective
tissue in which are
seen few cells, per-
meates the cortical
substance, being
more extensive in
the medullary por-
tions.

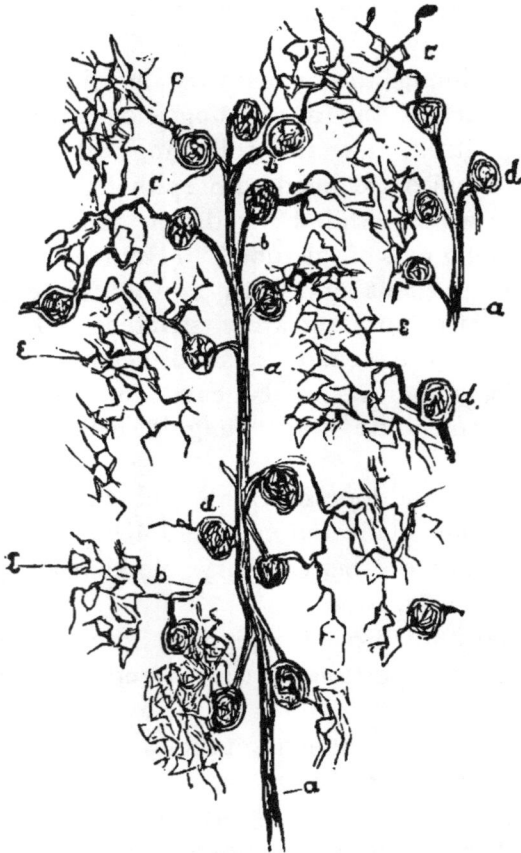

FIG. 106. Thin section of injected kidney of the pig. a,
artery. b, afferent and c, efferent vessels. d, glomeru . e,
capillaries. x 20.

The arterial sys-
tem is easily studied
in well injected specimens.   The renal artery enters at the
hilus and usually divides into four branches, which soon give
rise to other branches.   These pursue a straight course
through the medullary substance, and having reached the

boundary lines between the two substances, they bend over in such a way that the convexity of the arch looks towards the external cortical layer, while the concavity is toward the hilus. From the concave side, vessels are given off which are soon reduced to the size of capillaries, and which take a direction toward the papillæ. These are the *artcriolæ rectæ.*

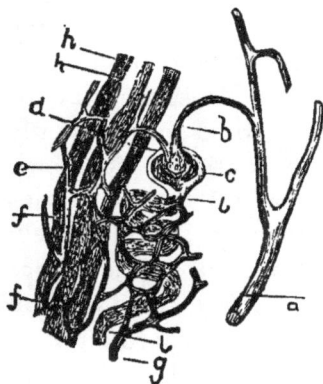

FIG. 107. From the kidney of the pig (semi-diagramatic). a, arterial branch. b, afferent vessels of the glomerulus, c. d, vas efferens. e, breaking up cf the same into the straight capillary plexus of the medullary ray. f, rounded plexus of the convoluted canals. i, g, commencement of the venous branch. (Frey).

From the convex surface a large number of branches arise nearly at right angles with the original branch. These branches pass straight through the centre of the cortical pyramid, as far as possible removed from the medullary rays. They are g'v'ng oℸ branches continually on either side, the vas afferens glomeruli which soon break up into capillaries without anastomoses. The venous system is more complicated. While, as a rule, there is but one efferent vessel to each Malpighian body, yet not unfrequently two, three, and rarely four are seen emerging from the capsule at a point corresponding to the entrance of the artery. These vessels form long, narrow meshes around the medullary rays, and narrow circular ones around the convoluted tubes. The majority of these capillaries are collected into venous branches which usually accompany the coil bearing artery. A system of capillaries seen in the medullary portion arises mainly from the dividing of efferent vessels, coming from Malpighian bodies lying deep in the kidney. They permeate the whole of the medullary substance and form a net-work around the openings of the collecting tubes on the papillæ.

Nerves and lymphatics are not wanting.

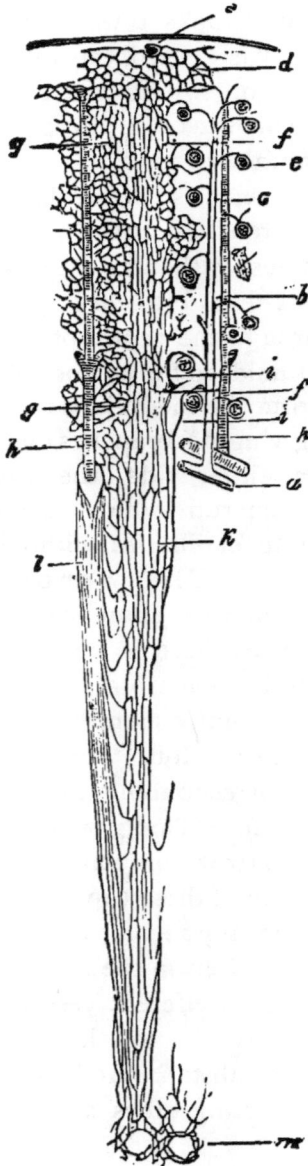

METHODS OF EXAMINING.

Sections of the fresh kidney may be made with a Valentin's knife, or a freezing microtome. The most instructive views are obtained only from hardened injected specimens. The kidney of the rabbit or pig may be injected with little trouble from the renal artery. Prussian blue or carmine will give good results if only the injection be continued for some time in order that all the capillaries of the glomeruli may be completely filled. The vein may be injected at the same time. After the injection the organ should be cut into eight or ten pieces, placed in Müller's fluid, and in two or three weeks transferred to alcohol, to complete the hardening. The sections should be cleared in oil of cloves, and mounted in dammar. If desired, they can be slightly stained in hæmatoxylin before clearing.

It is quite difficult to inject the kidney from the ureter, and if the injection be a success, the specimen will not show to any great advantage;

FIG. 108. The vascular arrangement of the kidney in vertical section. a, arterial branch at the margin between the cortex and medulla. b, coil-bearing artery. c, vasa afferentia of the glomeruli. d, capillary reticulum of the external cortical layer. e, vein of this part. f, elongated capillary net-work of the medullary rays. g, rounded net-work around the convoluted uriniferous canals of the cortical pyramids. h, venous branch of the cortex. i, efferent vessels of the deepest glomeruli. k, their capillary net-work. l, venous tubes of the medulla. m, capillary net-work of the papilla. (Ludwig.)

for owing to the convoluted course of the canals and their dense arrangement, it is quite impossible to trace the course of one tube for any considerable extent.   The tubes may be separated by teasing, if the interstitial connective tissue be previously destroyed.

This may be accomplished in a measure by boiling in acids, or by macerating the section for 12 or 14 hours in muriatic acid, to which has been added water till the acid has ceased to smoke.   It is then washed and placed in distilled water for a day, when the tubes may be isolated by careful teasing.   For isolating the tubes the following method of Henle's is quite successful : Sections of fresh kidney are placed in a flask partly filled with a mixture of eight parts of common alcohol and two parts hydrochloric acid.   The cork of the flask is pierced with a long glass tube. The specimen is boiled in this mixture for several hours, when the fluid is poured off and distilled water added.   After seven or ten days, the pieces can be carefully teased with needles.

The epithelium of the tubes is best studied in the kidney of the mature fœtus.   Small pieces are placed in a one per cent. solution of potassic bichromate for two weeks.   Thin sections are washed in water, and slowly stained with carmine.

CHAPTER XIV.

The Lymphatics.

THE lymphatic vessels are quite peculiar in this respect, that while they indirectly withdraw the fluid which they contain from the blood-vessels, they eventually return it to them by their terminal vessels. The lymphatic vessels in form and structure ordinarily agree with the blood-vessels, their walls being, perhaps, a trifle thinner and more transparent. Valves are met with similar to those of the veins. The very smallest lymphatics consist of a single layer of nucleated cell-plates like those of the capillary blood-vessels, and are brought to view by the staining action of a nitrate of silver solution. When it is considered how very delicate the walls of these capillaries are and how colorless their contents, it is understood why they are dissected with such great difficulty. For their examination artificial injections are requisite. Hyrtl came to our relief in offering the very effective method known as his "puncturing method." If a tissue is thought to contain lymphatics, the fine point of the syringe or injecting apparatus is carefully forced into the tissue and the mixture allowed to pass into it carefully and slowly in the hope that a wounded lymphatic may be injected. Several trials may be necessary but finally the object is accomplished. By staining the central tendon of the diaphragm with the silver nitrate solution openings of not inconsiderable size are seen between the epithelial cells. These openings—stomata—never exceed the size of an epithelial cell and are always as large as a red blood-corpuscle.

This rich plexus of openings is doubtless subservient to the absorption of the fluids of the peritoneal cavity.

FIG. 109. Central tendon of the diaphragm of the rabbit. a, lymphatic capillaries. b, serous canals. x 250.

Recklinghausen demonstrates these openings in the following manner : Milk is injected into the peritoneal cavity of a mammal (rabbit.) A cork ring is pressed against the central tendon from the thoracic side and needles are passed through the tissue into the cork ; in this way the surface of the tendon is procured in an absolutely uninjured state, and it can now be excised without difficulty or harm and transferred to the stage of the microscope. If a drop of milk be placed on the surface, the globules may be seen to enter the lymphatic vessels. The openings are so small that only two, or at most three globules can enter abreast. They are quite round and usually lead perpendicularly into the lymphatic vessels. Thus the great serous cavities of the body may be regarded as giving origin in some regions to the lymphatic vessels.

THE LYMPHATIC GLANDS.

The lymphatic glands appear oval or circular, or as is quite generally the case, bean-shaped. Each gland is surrounded by a sheath, which is composed of two strata of connective tissue. Direct prolongations of the inner stratum penetrate into the gland as membranous septa, radiating toward the point where the lymphatics emerge. These septa communicate freely with one another laterally, and having entered the gland for a certain distance they break up into trabeculæ which anastomose with one another to form small meshes. The part of the gland containing the septa is called the cortex, the part containing the trabeculæ, the medulla.

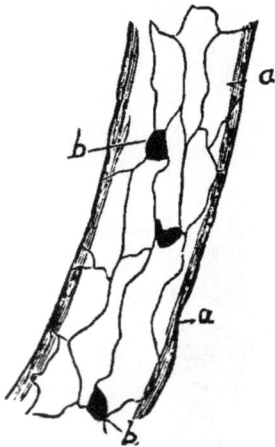

FIG. 110. Lymphatic canal. a, cell plates. b, spaces between the same. Silver staining. x 400.

The gland tissue, or the adenoid tissue of His, fills the meshes in the cortex and the medulla. In the cortex these meshes are oval and the enclosed gland tissue is called a follicle, and in the medulla, medullary cylinder. The follicles and medullary cylinders are never in close contact with the sheath and septa ; a space is always left, the lymph sinus of

His.   These sinuses, both in the medullary and cortical portions   intercommunicate.   The adenoid tissue has a like
structure wherever it is found.   There is a matrix of dense,

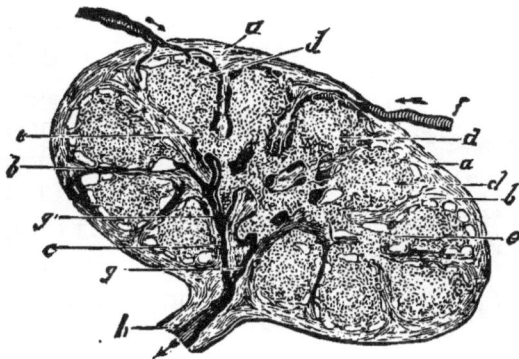

reticulated net-work
of fine fibrils, called
the adenoid reticu-
lum.   The endo-
thelial cells demon-
strated on this retic-
ulum   are   quite
readily removed by
pencilling.

The meshes of the
adenoid   reticulum
contain one or more
lymph   corpuscles,
which appear iden-
tical with the white
blood-corpuscles
with this difference :

FIG. 111. Section through one of the smaller lymphatic
glands, with the current of the lymph—half diagrammatic figure.
a, the capsule. b, septa between the follicles of the cortex (d)
c, system of septa of the medullary substance as far as the hilus
of the organ. e, lymph tubes of the medulla. f, lymphatic
passages, which surround the follicles and flow through the
spaces of the medulla. g, union of the latter into an afferent
vessel (h) at the hilus. (From Gray.)

their nuclei are much larger.   But believing the nucleus to
represent the living part of the cell, this is what we would expect to find.   When the corpuscle is young (lymph corpuscle)
it has a large nucleus, when it is older the nucleus is smaller
(white blood-corpuscle), and when quite old, the nucleus is
very small or has disappeared altogether (red blood-corpuscle).

The blood-vessels of the tissue enter the hilus and reach
the interior of the organ to terminate in a rounded capillary
net-work.

Upon entering a gland the lymph takes the following
course : From the afferent vessel it passes into the capsular
net-work, from here into the cortical and then into the medullary sinuses, from these into the net-work of the lymphatics at
the hilus and finally into the efferent trunk.   Besides these

there are various lymphoid organs in the body, either single or grouped, which are nearly identical with a lymphatic follicle. The following are among the number: the glands and follicles of the mucous membrane of the stomach and large and small intestines, the tonsils, the follicles of the conjunctiva, Peyer's glands, thymus gland, and also the spleen.

The tonsils are covered with pavement epithelium, beneath which is connective tissue containing oval or spherical lymph follicles. The thymus gland consists of masses of adenoid tissue anastomosing into a network.

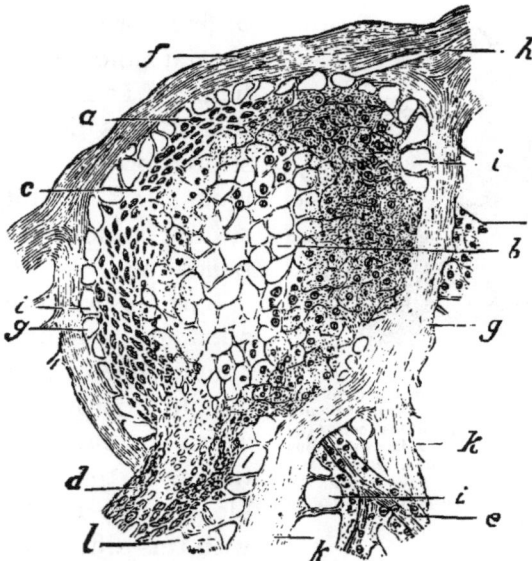

FIG. 112. Follicle from a lymphatic gland of the dog, in vertical section. a, reticular framework of the more external, b, of the internal portion. c, fine reticulum of the surface of the follicle. d, origin of a larger, and e, of a finer lymph tube. f, capsule. g, septa. k, division of the one. i, investment space and its tenter-fibres. h, vas afferens. l, attachment of the lymph tubes to the septa. (From Gray.)

Peyer's glands consist of a number of aggregated lymphoid follicles. In a single Peyer's patch may be fifteen or twenty of these follicles. They are often seen projecting into the lumen of the tube as convex bodies between adjacent villi or perhaps a few villi are scattered over them. The follicles are extraordinarily supplied with capillary blood-vessels which permeate them in a radial direction.

The vermiform process must be regarded as an enor-
mously developed Peyer's plate (Frey.)

THE SPLEEN.

The spleen is surrounded by two membranous coats, a
serous one derived from the peritoneum, and a stronger,
highly elastic fibrous one. This fibrous membrane sends septa
into the interior of the organ after the manner of the other

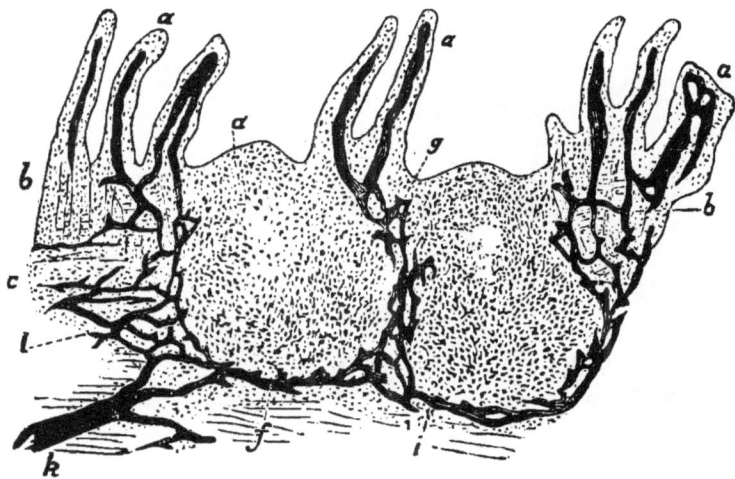

FIG. 113. Vertical section through a human Peyer's patch; a, intestinal villi. b, Lieberkühnian glands. c, muscular layer of the mucous membrane. d, apex of the follicle. f. basis
portion. g, lymph-passages around the follicle. i, at the base of tne same. k, lymphatics of the
sub-mucous tissue. l, lymphoid tissue of the latter. (After Frey.) x 32.

lymphatics. In the soft spleen substance are seen cylindrical
masses of adenoid tissue, averaging about the $\frac{1}{70}$ of an inch
in diameter, and surrounding the arterial branches. These
are the malpighian corpuscles or the lymphatic follicles of the
spleen.

In some cases these follicles are situated on the side of
the arterial wall, or the vessel may pass directly through their
centre, or what is most generally the case, the blood-vessels
are situated excentrically, surrounded by a greater amount of

adenoid tissue on one side than on the other. Branches extend into the corpuscle from the blood-vessel. Now the external connective-tissue coat of the small arteries becomes
transformed into lymphoid tissue, so that these Malpighian
corpuscles represent localized expansions of this external coat.
The meshes are filled with lymphoid cells, and capillaries are
freely supplied.

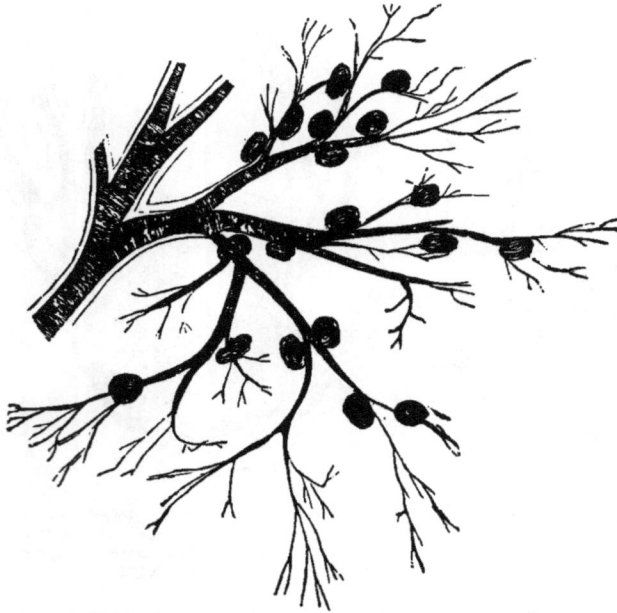

FIG. 114. Small artery, to which Malpighian corpuscles are attached. x 10.
(Kolliker.)

In the spleen pulp are a number of nucleated, branched
connective-tissue cells, often containing pigment granules in
their interior. These cells have no regular size or form, but
serve as a support to the soft pulp tissue. Besides these there
are red and white blood-corpuscles and also larger colorless
corpuscles in the interior of which are the remains of from one
to many red corpuscles (see figure 19.)

The small arteries terminate in capillaries, which, after a time, loose their cell walls, for, gradually, processes are seen extending from them to unite with the processes of the connective-tissue cells mentioned above. Here, then, is first a blood-vessel with loose branched cells for a wall, and finally this even disappears and we have but a channel for the blood in the soft pulp with no walls to confine it. The veins commence in the same way that the capillaries end. Thus in passing through the spleen the blood must come in immediate contact with the pulp tissue. (See origin and death of the red blood-corpuscle.)

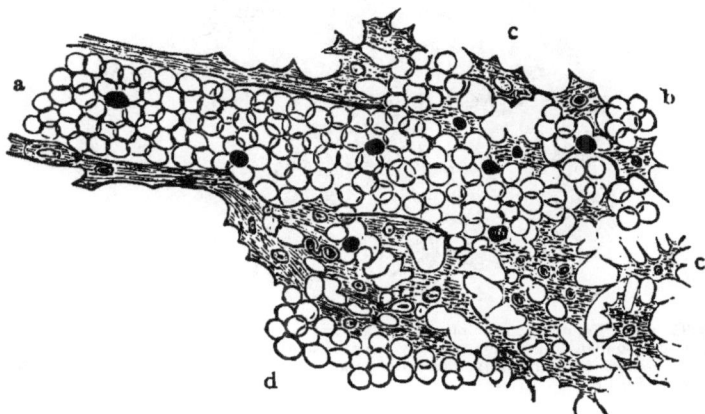

FIG. 115. Thin section of spleen pulp, showing the mode of origin of a small vein. Chromic acid preparation. a, the vein filled with blood-corpuscles. b, blood-corpuscles filling the interstices and in continuity with a. c, branched cells. The shaded bodies in the vein are the white corpuscles. (Quain.)

LYMPH CAPILLARIES.

The lymph capillaries are related to the tissue in which they originate as follows : They commence in a system of lacunæ which are united with each other by larger or smaller canals. This is the lymph-canalicular system of von Recklinghausen. The spaces are filled with a soft, semi-fluid substance in which are found the migratory connective-ssitue cells ; the stable cells, uniting with their processes to form the walls of

these spaces ; they are continuous with the endothelial wall of
the lymph-capillaries.   This lymph-canalicular system is very
generally distributed in the different organs of the body, dif-
fering but slightly wherever found, owing to the arrangement
of the matrix of the part.   In the lungs, bone, cornea, serous
and synovial membranes, the system consists of lacunæ, lined
with cells as described above.   In tendon, fascia, muscle and
nerve, it consists of long straight channels between the connective-tissue bundles and fibres.   In the skin and mucous
membranes it consists of inter-fascicular spaces, irregular in

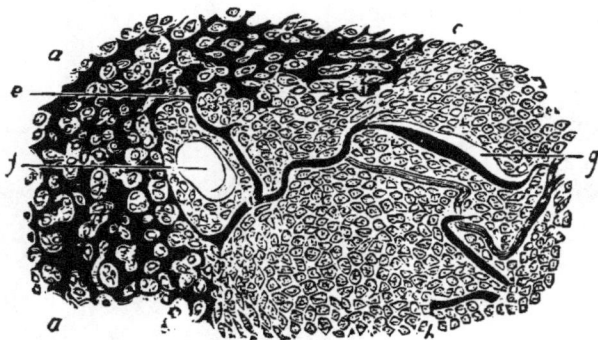

FIG. 116. From the spleen of the hedgehog; a, pulp, with the intermediate currents.
b, follicle. c, boundary layer of the same. g, its capillaries. e, transition of the same into the intermediate pulp-current. f, transverse section of an arterial branch, at the border of the Malpighian
corpuscles. (Frey.)

shape and situated between bundles of fibres as they cross and
recross each other.   It will be remembered that the walls of
the blood-capillaries permit particles to pass through their
stomata into the surrounding connective-tissue.   Such particles pass at once into the lacunæ, lined by the loose,
branched connective-tissue cells ; from this lymph-canalicular
system they pass very readily into the lymph capillaries.

The reasons why a current should be directed from the
blood-vessels to the lymphatics are these : first, the pressure
in the blood-vessels is considerable while in the lymphatics it
is nothing ; and second, the blood-vessels are in the centre of

an area while the lymphatics are at the periphery, hence the tendency of the lymph to travel the intervening systems (von Recklinghausen.)

In inflammation it is well illustrated how formed matter passes from the blood-vessels into the lymph-canalicular system and from here to the lymphatics (Cohnheim and others.) Arnold introduced Berlin blue into the blood-vessels of living rabbits and frogs, and afterwards traced the pigments through the lymph-canalicular system directly into the lymphatics.

METHODS OF EXAMINING.

For the injection of the lymph sinuses the puncturing method of Ludwig should be followed. This consists in thrusting a fine needle point of the syringe into the tissue anywhere and in forcing the injecting fluid into it, letting it go wherever it will. To be successful this method requires a large experience and the student must expect a good proportion of failures at the start. After injection the tissue is hardened in alcohol. Nitrate of silver used as an injection or external application demonstrates the cellular structure of the lymphatic canals. This method is easily carried out by using the mesentery of some small animal, as the cat or rabbit ; a piece of the mesentery is stretched on cork, the endothelial cells removed by pencilling, the silver solution applied, and the specimen treated as recommended for all silver staining.

The glands of some of the lower animals, dog, cat, are hardened in alcohol, and the thin sections slightly stained. Glands placed first in Müller's fluid and subsequently in alcohol harden well and are very suitable for section-making. To examine the framework the sections should be very thin and thoroughly pencilled.

A natural injection of the lacteals is easily procured by killing an animal, as a cat, about four or five hours after a hearty meal of milk. The entire lacteal system will be found fully distended with the digested fat.

Sections of Peyer's patches, with injected blood-vessels are readily made after alcohol hardening.

The lymph passages of the mesenteric glands of the ox can be beautifully injected by the puncturing method.

CHAPTER XV.

Nerve Fibres and their Modes of Termination.

A SIMPLE classification of nerve fibres can be made by dividing them into two general classes. First, those having the axis cylinder surrounded by a medullary substance, and second, those consisting of the axis and sheath alone.

MEDULLATED FIBRES.

These are by far the most generally distributed. They present the greatest variation in size; some are as large as the $\frac{1}{1200}$ of an inch, while others are not half the size, and if examined near their origin or termination they may be no larger than the $\frac{1}{14000}$ of an inch. Every medullated nerve fibre is now known to consist of three parts; 1, the sheath; 2, the medullary substance; 3, the nerve axis. The investing membrane consists of a fine, transparent, homogeneous, connective-tissue, known as "the sheath of Schwann," "limiting membrane of Valentin," "primitive sheath," "neurilemma." It corresponds to the sarcolemma of muscle. It covers the medullary substance wherever it exists and forms a protecting envelope to the delicate nerve axis throughout its whole length, save at its very origin and termination. It is not seen without the use of reagents. Here and there at quite regular intervals is a nucleus.

The medullary substance completely surrounds the nerve axis and occupies all the space between it and the neurilemma. It is known as "the white substance of Schwann," "the myeline," "nerve medulla," etc.

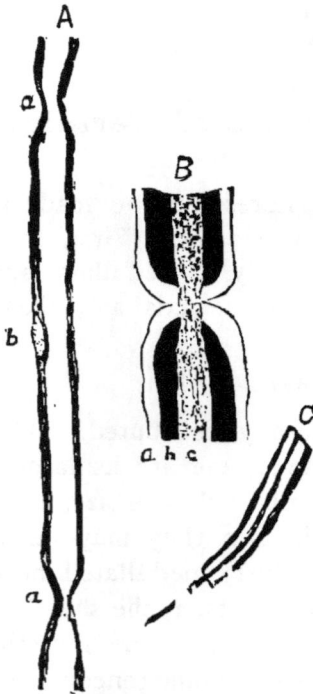

FIG. 117. Medullated nerve fibres. A, from frog, fresh specimen. a, nodes of Ranvier. b. nucleus. B, node of Ranvier, osmic acid preparation. a, sheath of Schwann. b, medullary substance. c, axis cylinder. C, nerve fibre treated with alcohol, the axis cylinder is protruding from the sheath. A and C x400. B x750.

It is composed of a peculiar combination of albuminoid bodies, is homogeneous and transparent when fresh, and is capable of greatly refracting the light. If the nerve be allowed to dry, or if one of many reagents be allowed to come in contact with it, the medullary substance coagulates and becomes very opaque, completely hiding the axis from view. This gives a very peculiar and characteristic appearance, and the fibre is said to become "varicose."

The axis cylinder is seen in a transverse section of a bundle of nerves as a small cylindrical body occupying from one-fifth to one-fourth the diameter of the entire fibre. It is the sole essential constituent of the nerve. It exists alone at the very commencement and termination of all nerve fibres. It is not readily seen in the fresh nerve without reagents, but is very easily discerned in stained transverse sections of the spinal chord. Besides pale granules in the axis, certain reagents bring out well-marked longitudinal striæ. They are compared to the longitudinal striæ of muscle, dividing the nerve, as the muscle, into the " primitive fibrillæ " of Schultz, or into the "axis fibrillæ" of Waldeyer. After the study of the modes of termination of these fibres, additional proof will be seen that each nerve axis is a bundle of most delicate fibrils.

Situated at quite regular intervals on these fibres are constrictions, which, for a long time, were supposed to be due to the methods of manipulation. Ranvier has shown these "constriction rings" to exist in many different animals, and that about midway between two of these constrictions is a nucleated nerve corpuscle. Thus it is in mammals, birds and amphibia, while in fishes the number of nuclei is greater. This makes a nerve fibre composed of a number of long narrow cells placed end to end, each cell possessed with one nucleus, or in the fishes with many nuclei. Here, then, at each constriction ring, nutrient matter may find a place to enter the highly endowed axis, and the results of decomposition find an exit. The nerves of the body consist of a greater or less number of nerve fibres united together by a connective tissue, called the perineurium, which corresponds to the perimyseum of muscle.

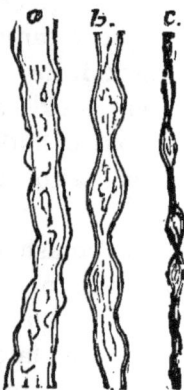

FiG. 118. Varicose nerve fibres. c. from the brain. x 400.

METHODS OF EXAMINING.

Cut out a small piece from the sciatic of the frog and place on the glass slide, add a drop of chloroform and tease carefully. Then add more chloroform and cover with thin glass. If the chloroform evaporates add more. This reagent dissolves the oily medullary substance and brings the axis cylinder into view.

Prepare another specimen and add strong acetic acid.

The substance of the fibres will retract except the axis, which will protrude from the end sufficiently to be studied.

A piece of a small nerve is carefully cut out and placed in a 1 per cent. solution of osmic acid. Care should be taken not to injure the nerve in any way. Let it remain in this reagent about four hours, or until it is stained black. Then wash thoroughly in water and tease in dilute glycerine. Add hæmatoxylin and let it remain on the specimen until it is colored well. Wash in dilute acetic acid and examine in glycerine. Unless stained too deeply, all parts of the fibre can be made out. The medullary sheath will appear black from the osmic acid staining, and the constriction rings will look like breaks in the deep coloring. At intervals the axis cylinder will be seen stained with the hæmatoxylin, as will also the nuclei of the neurilemma.

Take a small piece of nerve and add caustic potash or soda. This will render more fluid the nerve-medulla, and if pressure be applied to the cover-glass, it may be forced out of the tubes in large fat-like drops. Then the primitive sheath may be seen. By teasing a nerve-trunk, many times the needles will cause a displacement of the contents of the sheath, so that the latter is brought to view for a short distance, much as the sarcolemma is seen under like circumstances. Silver staining will demonstrate the nuclei of the sheath to good advantage. A 5 p. c. solution should be used and the fibres allowed to remain in it only two or three minutes when they are treated after the usual fashion for silver staining.

FIG. 119. Axis cylinder, nitrate of silver preparation, showing longitudinal striation. x 1000.

NON-MEDULLATED FIBRES.

Early in the history of the fœtus the nerve fibres of the en-

tire system are of this class.   The olfactory nerve of man and
of all   vertebrates is composed exclusively of these fibres.

They also form a large part of the sympathetic system. These fibres consist of
an axis cylinder closely surrounded by a
nucleated sheath.   Remak discovered
them first as composing entirely the thick
splenic nerves of ruminants.   For this
reason, these fibres are often called
" Remak's fibres."   Remak has named
them "ganglionic fibres."   They are
most generally found in the invertebrata.
Under the microscope they are not so
bright as the other class, but look gray
and gelatinous.   The neurilemma is very

FIG   120.   Non-medullated
nerve fibres from sympathetic.
x 400.

firmly united to the axis, and is of firm consistence itself.
For a bundle of these fibres can be teased apart without harm.

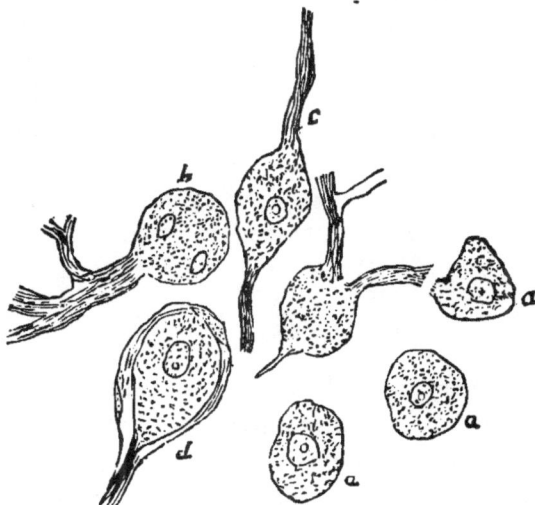

FIG. 121. Ganglia cells. a, apolar. b, unipolar. c, bipolar. d, unipolar cell, with
nucleated cell capsule which is continued over cell process as the sheath of Schwann.
x 250.

These are procured for study from the cervical sympathetic of an animal, or from the sympathetic nerves of the frog, which lie along the vertebral column in close connection with the aorta.   They are treated as described for medullated fibres, although simple hæmatoxylin staining answers well, staining the nuclei in a beautiful manner.

CEREBRO-SPINAL GANGLIA.

The frame-work of these ganglia is of connective-tissue, supplied by the dura-mater and arachnoid, and is supplied freely with blood-vessels.   The cells are spherical or irregularly oval, and are of all sizes.   Each cell is provided with a nucleus, and many times a nucleolus.   The majority of writers give to these cells but one process.   As this process is directly connected with a nerve axis, it is called the axis cylinder process.   The substance of the cell being composed of minute fibres, intra-cellular fibrils, this process must be considered as a prolongation of these.   Some cells do not have even one process; they are apolar.   Each ganglion cell is surrounded closely by a membrane lined with nucleated cells; these are continued along the axis cylinder process to make the nerve corpuscles beneath the neurilemma.   The membrane itself extends over the axis cylinder as the sheath of Schwann.   A medullary substance may or may not be supplied.

SYMPATHETIC GANGLIA.

These are very like those of the cerebro-spinal system with this distinction.   Cells are present here from the apolar to the multipolar.   No matter how many processes may be given off, each one is a continuation of the cell substance, and receives a prolongation from its capsule with its lining cells.

MODES OF TERMINATION.

An interesting question arises concerning the modes of termination of the nerve fibres in the different tissues of the body. First of all must be considered how the nerves divide on their way to the periphery. When near its peripheric extremity the fibre divides usually into two branches, each of which may divide again and again, until from ten to thirty or more branches arise from the one, or one fibre may suddenly break up into a number of branches at once. Brücke and Müller were the first to observe these modes of division. Belharz has told us that in the electric eel a single medullated fibre extends from the medulla oblongata to each of the two electric organs, and in order that each electric plate

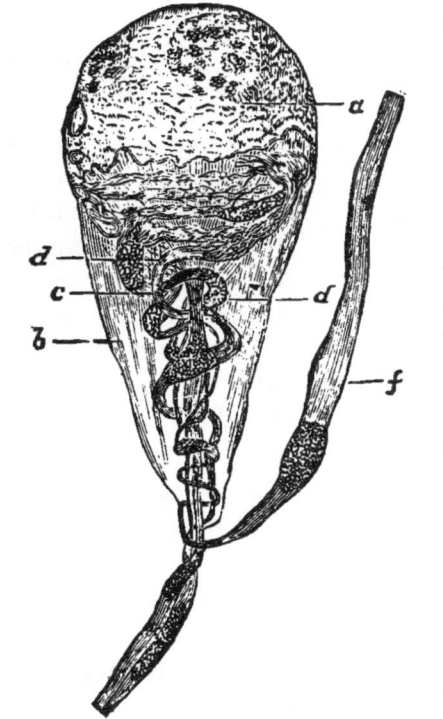

FIG. 122. Ganglion cell from the sympathetic of the hyla, or green-tree frog. a, cell body. b, sheath. c, straight nerve fibre. d, spiral fibre. continuation of the former, e ; and of the latter, f.

may have a terminal fibre, it must divide millions of times.

As a result of each of the divisions, the axis cylinder becomes proportionately smaller, so that the sum of all the nerve axes of the terminal branches of any given fibre will exactly equal the original axis. This is not true of either the medullary substance or sheath of Schwann ; these increase soon after division, until each of the branches is nearly as

well supplied with them as was the original fibre.   Both the
neurilemma and medullary substance disappear sooner or
later, and are absent
in the ultimate fibrils.

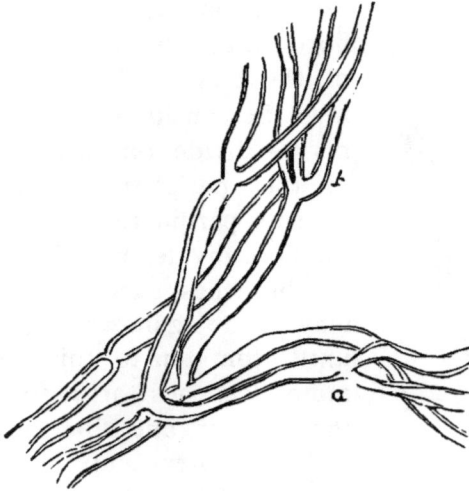

TERMINATION IN STRIPED MUSCLE.

When one of these
nerve branches
reaches a muscle fibre
the following changes
occur: The neuril-
emma of the nerve
becomes continuous
with the sarcolemma
of the muscle, and
the medullary sub-
stance ceases, while
the axis appears to

FIG. 123. Muscular nerves of the frog, showing division of
the fibres. a, into two, b, into three branches. x350. (After
Kölliker.

rest on a peculiar, nucleated, rounded plate.   This plate is
concave within and convex without to conform with the muscle
fibre.   On it are from four to twenty nuclei.   This is the terminal plate of Krause, or the nerve end-plate of Kuehne.
Viewed in profile this end-plate presents a regular convex
elevation—"Doyère's mount."   As soon as the axis reaches
this end-plate, it breaks up into a number of antler-like
branches, or into several minute fibres, which form a net-
work with one another.   Into some of these end-plates two
nerve fibres enter.   They differ in size, and many muscle
fibres have more than one of them.

IN UNSTRIPED MUSCLE.

When a nerve fibre reaches the unstriated muscle in which
it is to terminate, it divides into a number of non-medullated

branches. The axis cylinders of these branches now divide and collect in small groups, and by a re-arrangement of their fibrils form a plexus—"the ground plexus of Arnold."

Branches of this plexus divide into still smaller branches to form "the intermediary plexus of Arnold." These branches are distributed to the separate bundles of unstriped fibres. Although these branches are so very small, yet they are composed of a number of primitive fibrils. They now break up into the smallest branches which penetrate the substance that lies between the individual cells. A net-work is formed here by anastomosing branches. These are the "muscular fibrils of Kebs." Farther than this, it is difficult to trace these fibrils. We have never been able to see anything like the terminations to be described.

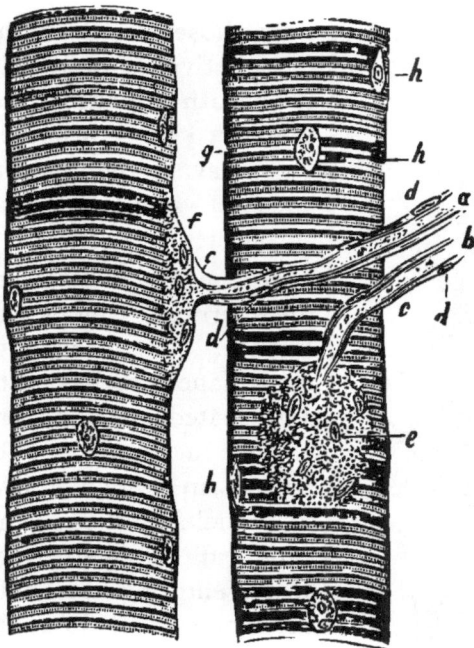

FIG. 124. Two muscular filaments from the psoas of the Guinea-pig, with the nerve terminations; a, b, the primitive fibres and their continuation into the two terminal plates e, f c, neurilemma with nuclei. d. d, and passing over into the sarcolemma. g, g. h, muscular nuclei. (Frey.)

Frankenhäusen describes still finer branchlets given off from this net-work, which terminate in the nucleolus of each cell.

Elischer says they terminate in a blunt, enlarged point on the surface of the nucleus. Klein thinks their termination in the nucleus not improbable.

IN TENDONS.

Here the nerve axis divides into fine branches, which again divide into the most minute elementary fibrils, to form a close net-work. The terminal end is embedded sometimes in a nucleated substance something like the end-plate in muscle. (Rollett.)

Bulbs are seen in some tendons like those found in the conjunctiva. (Golgi). In the sheath of tendon one observer has described bulbs having the appearance of small Pacinian corpuscles.

In the conjunctiva are the terminal bulbs of Krause. These are elliptical-shaped bodies with a nucleated envelope, enclosing a hyaline substance, in which are a few nuclei, although no cells can be perceived. Upon entering this bulb the nerve fibre loses its medullary substance, and the axis passes through the substance to the opposite pole and terminates in a slight enlargement. They are not easily seen, and their existence in man is denied by some investigators. Nothing can be said concerning them from personal observation.

FIG. 125. Two muscle fibres from hyo-glossus of frog. a, nerve end-plates. b, nerve fibres leaving the end-plate. c, nerve fibres terminating after dividing into several branches. d, a nucleus in which two nerve fibres anastomose. x 600 (Klein and Smith.

The Pacinian bodies were described as long ago as 1741, but were soon forgotten to be rediscovered in 1830 by Pacini. These are elliptical bodies averaging $\frac{1}{12}$ of an inch, and present to the touch a firm, tense structure. In man they are found in the palm of the hand and in the sole of the feet. The mesentery of the cat is an admirable place to procure them, many times in great quantities.   The capsule is marked by concentric striations, which in turn denote their connective-tissue membranes.   On the inner surface of this capsule

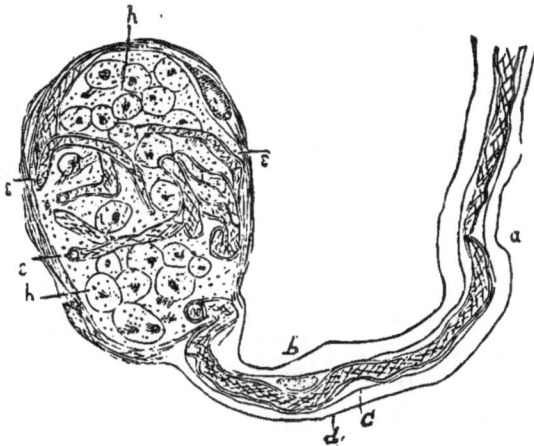

FIG. 126. End bulb of the conjunctiva of man.   a, node of Ranvier. b, nerve nucleus. c, sheath of Schwann. d, sheath of Henle. e, branches of the axis cylinder. h, granular substance with nuclei.   x 700 (Key and Retzius.)

there is a single layer of flat cells.   When the nerve fibre enters this body it loses its medullary substance, the neurilemma becomes continuous with the capsule, and the axis cylinder enters alone, terminating in a divided extremity, or undivided in a pear-shaped swelling.

The axis cylinder is striated most beautifully in a longitudinal manner, showing it to consist of primitive fibrillæ. Around this terminal fibre is a transparent, striated, semi-fluid central matrix.   Between the layers of the capsule are capil-

FIG. 127. Pacinian corpuscle from mesentery of the cat. x 35.

lary blood-vessels. Specimens are very frequently obtained showing a capillary entering the corpuscle with the nerve.

In the skin there are the tactile corpuscles of Meissner and terminal nerve fibres. The tactile corpuscles are most numerous on the palmar surfaces of the hands and fingers, and plantar surfaces of the feet and toes. On the third phalanx of the index finger Meissner counted one hundred and eight in a space $\frac{1}{80}$ of an inch square. They are oval-shaped bodies about $\frac{1}{350}$ of an inch long. In the capsule are longitudinally and transversely arranged nuclei. The nerve axis alone enters this body and terminates in curved or looped windings. It is difficult to decide precisely how these fibres e n d, w h e t h e r i n a branched extremity or in a bulbous enlargement. Many times several fibres or pale axes are seen pursuing an oblique or longitudinal direction in the central part of the corpuscle.

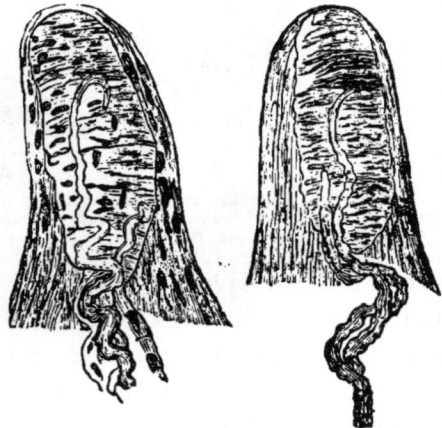

FIG. 128. Two human nervous papillæ from the skin of the volar surface of the index finger. In the interior of the papillæ is the tactile body, into the tissues of which the nerve fibres enter. (After Kölliker.)

Other nerve fibres, after forming a plexus just beneath

the rete-mucosum, give off branches which form a minute fibrillar net-work just beneath the epithelial layer.

In mucous membranes this fibrillar net-work probably exists. In the cornea are two systems. The deep system terminates in the true corneal tissue as a net-work, while from this a superficial net-work arises, which penetrates the covering of epithelial cells, and terminates just beneath the superficial layer of flattened cells either as a fine net-werk or in blunt ends. The terminal fibres do not enter into any relation with the cells themselves (Klein). For the termination of dental nerves see chapter on teeth. For the terminations of nerve fibres of the various organs of the body, consult the articles on those subjects.

CHAPTER XVI.

Spinal Cord.

THE spinal cord is surrounded by the dura mater, arachnoidea, and pia mater. Between the last two is a spongy tissue, consisting of smaller and larger trabeculæ of connective-tissue. This is the subarachnoidean tissue.

The ligamentum denticulatum extends between these two membranes the whole length of the cord. It is situated midway between the anterior (ventral) and posterior (dorsal) nerve roots dividing the subarachnoidean space into anterior and posterior spaces.

The spinal cord is the cylindrical elongated cord contained in the spinal canal. In length it is from sixteen to eighteen inches, and devoid of its membranes its weight is about an ounce and a half. It is a column of nervous tissue commencing above at the foramen magnum and terminating below on a level with the first lumbar vertebra in a conical filament of gray matter termed the filum termi-

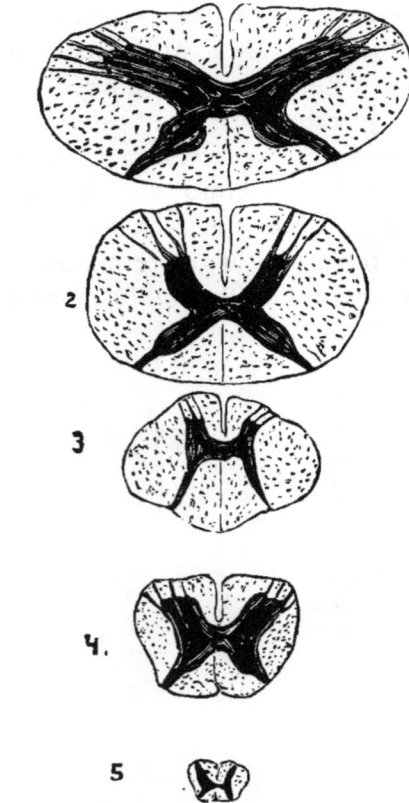

FIG. 129. Transverse sections of the spinal cord of 1, the horse. 2, the ox. 3, man. 4, the dog. 5, the frog. All x 2.

186

nale. It is marked by two enlargements, the one commencing at the third cervical and extending to the first dorsal, called the cervical enlargement, and corresponding to the origin of the nerves that supply the upper extremities. The other is opposite the last dorsal vertebra and corresponds to the origin of the nerves supplying the lower extremities, known as the lumbar enlargement.

A longitudinal fissure extends along the whole length of the cord in the anterior median line—the anterior median fissure. A like fissure extends along the posterior median line—the posterior median fissure. This divides the cord into two equal parts, which are united in the centre by a transverse band of nervous tissue. The anterior median fissure is shallower and wider than the posterior, and does not extend to the centre of the cord. It reaches to about one-third the thickness of the cord and is lined completely by a reflection of the pia mater.

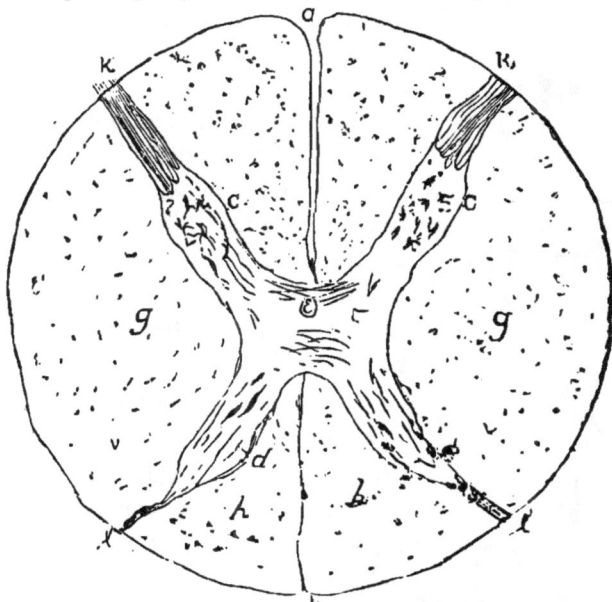

FIG. 130. Diagram of transverse section of the spinal cord. a, anterior. b, posterior median fissure. c, anterior, d, posterior cornua, e, central canal. f, anterior, g, middle, h, posterior columns.; k, anterior, l, posterior, nerve-roots.

The posterior fissure is not so distinct as the anterior, but extends nearly to the centre, this also receives a fold of the pia mater. The two surfaces of this reflected fold often become united so that the fissure is obliterated.

FRAME-WORK OF THE CORD.

Prolongations of the pia mater not only enter the fissures

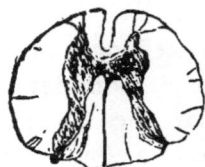
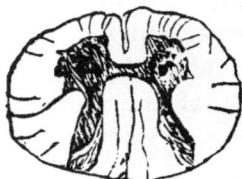
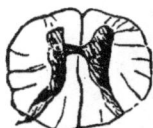

but also minute septa between sections of the white substance. They carry blood-vessels with them. A semifluid, homogeneous substance fills all the interstices between the nerve substance proper of the cord. This is the neuroglia matrix of Klein. There are also minute fibrils, forming a net-work, quite similar to elastic tissue, neuroglia fibrils. These fibrils pursue a longitudinal direction in the white substance, but every direction in the gray. There are present also branched, nucleated, connective-tissue cells. The "neuroglia" of the cord is composed of all three, matrix, fibrils, and cells. The greater the one of these parts, so much the greater the others. The amount of neuroglia varies in the different parts of the cord.

FIG. 131. Transverse section of human spinal cord at different heights. A, upper cervical. B, cervical enlargement. C, dorsal. D lumbar enlargement. E, sacral. F, coccygeal, (partly from Quain). x 2.

WHITE SUBSTANCE.

A thin transverse section of the cord reveals two substances, a central or gray substance, resembling in shape a capital H, and a peripheral or white substance. The white substance of each half of the cord may be divided into three columns. The ante-

rior (ventral) column includes all that part of the cord which
is bounded internally by the anterior median fissure, externally by the anterior cornu and the nerves emerging from
it to give rise to the anterior spinal nerves, and posteriorly
by the gray substance.   It is continuous with the anterior
pyramids of the medulla.   The lateral column is bounded

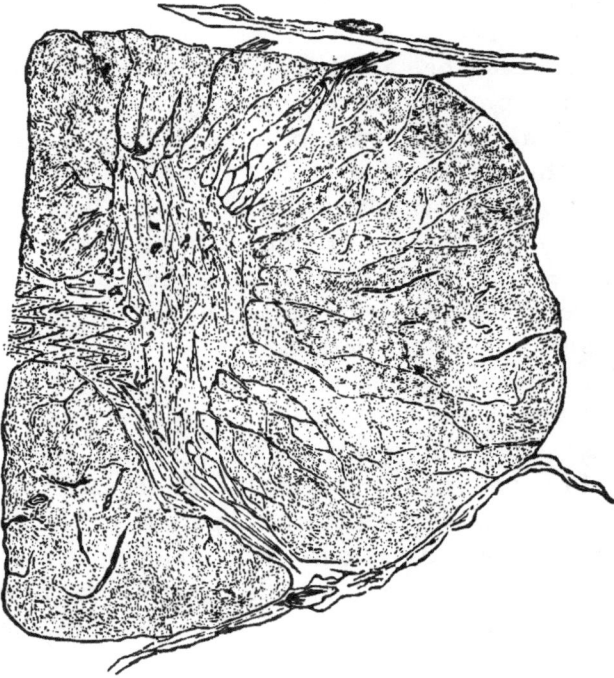

FIG. 132. Transverse section of one-half of the spinal cord of a dog. x 15.

internally by the gray matter, and anteriorly and posteriorly
by the anterior and posterior cornua.   It is continuous with
the lateral column of the medulla, and is much the largest
of the three columns.

The posterior (dorsal) column is bounded internally by the posterior median fissure and externally by the posterior cornu. This is directly continuous with the restiform body of the medulla.

Besides the neuroglia and its nutritive system of vessels, nerve fibres are seen running mostly in a longitudinal direction. In a transverse section of the cord the white substance appears composed entirely of these minute cylinders cut across. The axes of these fibres take the staining readily, and each one is seen surrounded by a transparent zone which represents the myeline. Just external to this is the frame-work of the cord, for there is no positive evidence of a sheath of Schwann. Not all of these fibres, however, pursue a horizontal direction, for in a transverse section, just at the base of the anterior median fissure, a band of nerve fibres is seen passing horizontally from the gray matter of one side to the white of the other ; this is the anterior white commissure.

This commissure makes a direct connection between the white matter of one side and the gray matter of the opposite side. Still other fibres are seen passing from the gray matter of the anterior cornua into the white matter of the lateral tracts, also similar fibres from the posterior cornua pursuing the same oblique direction. Horizontal fibres pass from the posterior nerve roots into the gray matter of the posterior cornua, and some of their fibres having their origin in the anterior cornua pass obliquely through the white substance to form the anterior nerve roots.

THE GRAY SUBSTANCE

is represented by the Latin capital H, and presents two anterior and two posterior cornua. The anterior cornua are thicker, blunter, with a serrated appearance at the margin, and they do not extend to the surface of the cord. Here are found the largest nerve cells in the body. In a well stained specimen of the spinal cord of the dog in the author's possession these large cells are visible to the naked eye. They are arranged in three groups, inner, anterior, and lateral, named in the order of the size of their cells. The cells are large, with branched and

FIG. 133. Nerve cells from anterior horn of the spinal cord of the ox. Obtained from fresh specimen by staining with hæmatoxylin. x 150.

unbranched procesess, having a nucleus and nucleolus surrounded by a membrane with an intranuclear net-work (Klein). The unbranched processes are known as "axis-cylinder" prolongations. While there is usually one, there may be two of these in one cell. They ultimately receive a medullary substance and sheath to form a nerve fibre. Cells are known as apolar, unipolar, bipolar and multipolar, according to the number of processes or poles proceeding from them.   Small nerve cells are found here which exist in greater numbers in the posterior cornua.

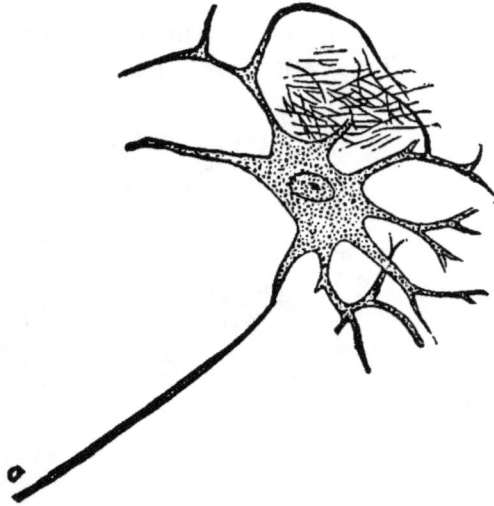

FIG. 134.   Nerve cell from anterior horn of the human cord. a, axis cylinder process. (Gerlach).   x 150.

The posterior cornua are narrower and longer, extending nearer to the surface than the anterior.   Near their termination is an enlargement, named from its appearance, substantia gelatinosa.   At the base of the posterior cornua, internally, a little back of the central canal is a mass of cells, restricted nearly to the dorsal region, known as Clarke's column.

A little anterior to the middle of the commissure that connects the two parts of the cord is seen the central canal which varies in diameter throughout the whole length of the cord.   In carefully prepared specimens

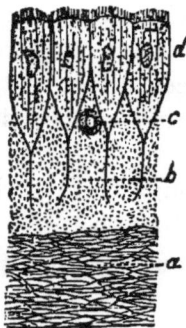

FIG. 135. Wall of the central canal of the spinal cord. a, connective tissue. b, fine granular substance. c, ciliated cell in way of development. d, ciliated epithelium. (Gerlach.)

it is visible to the naked eye. It is in communication above with the fourth ventricle and extends below to the ter. mination of the cord. It is lined with ciliated, columnar, epithelial cells which have a long, slender filament extending into the connective-tissue, which surrounds the canal just outside the cells. The space between the filaments of these cells and the connective-tissue is filled with a fine granular substance, which Gerlach believes to be "connective-tissue, devoid of elastic fibre net-work." A few nuclei are seen here, from which probably will be developed new epithelial cells. In early life this canal is filled with a fluid,—the cerebro-spinal fluid—but in the adult the canal is more or less compressed by the proliferation of the cells, and not capable of holding so much fluid. Anterior to the central canal, between it and the anterior white commissure, is a band of gray matter, the anterior gray commissure. Posterior to the central canal, between it and the base of the posterior median fissure is a similar band of gray matter, the posterior gray commissure.

Beside the cells in the gray matter, there is a minute net-work of fibrils known as "Gerlach's nerve net-work." This net-work is composed of minute primitive nerve fibrils, some of which have been seen to anastomose with the ultimate branches or divisions of nerve fibres from the posterior nerve root. All the branched processes of the nerve cells anastomose with Gerlach's nerve net-work. Nerve cells in the anterior cornua are attached to this net-work, while at the same time they are attached, through their axis-cylinder prolongations to medullated nerve fibres. Although the nerve cells in the posterior cornua anastomose with Gerlach's net-work, there is no direct union with nerve fibres.

Some of the nerve fibres originating in the anterior cornua pass to the lateral tracts and proceed direct to the brain as longitudinal fibres, not attached to any ganglion cells. (Klein).

Numerous fibres pass horizontally through the anterior portion of the posterior horns. (Gerlach. Brauch).

Bundles of fibres run longitudinally through about the middle of the posterior horns (Smith).

In the dorsal region are bundles of fine fibres from Clarke's columns of ganglion cells running in three directions, 1, backward, 2, crossing each other, and 3, passing in an outward direction (Gerlach).

The spinal cord is well supplied with blood-vessels.

METHODS OF EXAMINING.

Many methods are given for preparing the cord for study. The following method is in constant use at this laboratory: The cord is removed carefully from its bed and cut at once, with a razor wet with alcohol, into pieces about ¾ of an inch in length. These are immediately placed in a large amount of Müller's fluid. In a few days they are taken out and stripped of their membranes. This is not done at first from fear of injuring the soft tissue. They are replaced in the fluid where they can remain for an indefinite length of time without injury, or in a few days more may be placed in alcohol where they remain for three or four days, when they are transferred to absolute alcohol, until the requisite hardness is obtained. This is discovered by repeated trials although it will not vary far from twenty-four to thirty-six hours. If over-hardened the sections crumble ; if not hardened sufficiently the tissue springs, and only thick sections can be obtained. The piece is usually embedded, and then either held in the hand or placed in the microtome, the tissue and razor being flooded with alcohol. The thin section is now slightly washed by allowing a few drops of water to flow over it. Hæmatoxylin, carmine and aniline blue are valuable coloring agents. When carmine is used, it is better to dilute Beal's carmine one-half with water and to allow the sections to remain in it from two to twelve hours. For class work, it is used full strength, and the specimens remain in it from fifteen to thirty minutes. Distilled water is now flowed over the section until the excess of carmine is removed, when a few drops of a one per cent. solution of acetic acid are added. This is removed and alcohol substituted, which is replaced by absolute alcohol in about fifteen minutes. In about five minutes this is removed and the oil of cloves added. In a short time the specimens have cleared, when they are preserved in dammar. This gives great contrast between the two substances of the cord, stains the cells, their nucl i and nucleoli and the axis-cylinders in a most beautiful manner. Fresh specimens may be placed in the freezing microtome, the sections stained, and satisfactorily examined immediately. There are many valuable methods of treating the cord for examination; these are given in full in our works on microscopical technology.

CHAPTER XVII.

The Brain.

THE neuroglia of the brain is quite similar to that of the spinal cord. There are a few differences, *e. g.* in the white matter are seen small round cells between bundles of fibres, which are collected in masses in certain parts of the brain, as in the olfactory lobes. Boll describes connective-tissue cells in the gray matter surrounding the blood-vessels. Duke Charles, of Bavaria, says that there are colorless blood-corpuscles around these vessels both in disease and in health. The membrane lining the ventricles of the brain is a continuation of that lining the central canal of the spinal cord and is covered with a similar layer of ciliated cells. It is an accumulation of neuroglia, and is known as the ependyma.

WHITE MATTER.

In the white substance of the brain the nerve fibres are medullated as in the spinal cord, but are without any sheath of Schwann. Some of these fibres are extremely minute, while others are of medium size. These fibres connect the different parts of the brain substance as follows:

1. Those uniting the gray matter of the hemispheres with the large cerebral ganglia.

2. Those uniting identical parts of the two hemispheres *e. g.* corpus callosum.

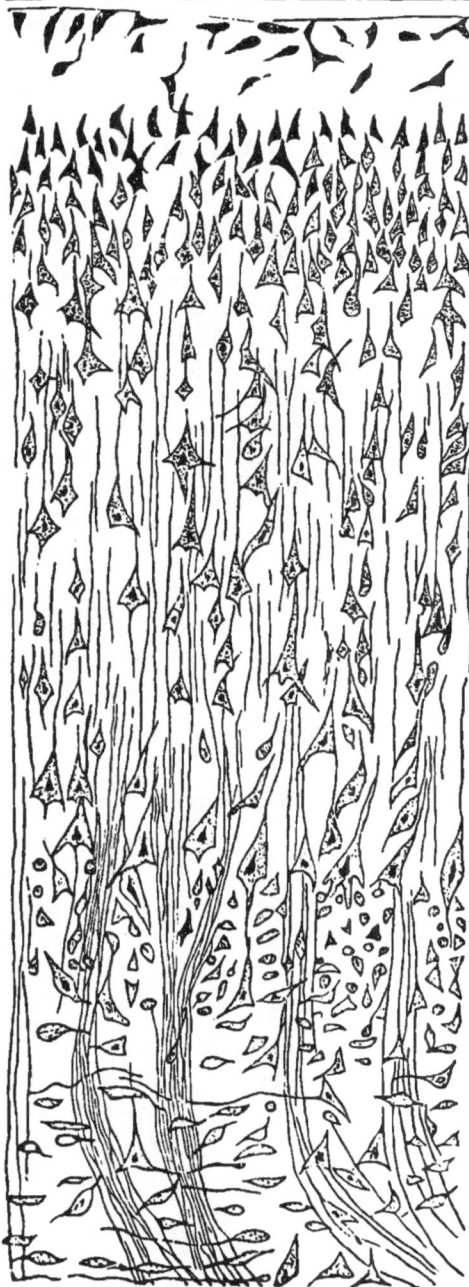

3. Those joining different parts of the same hemsiphere.

4. Bundles of fibres connecting the hemispheres with the cerebellum.

We owe our present knowledge largely to Meynert, the highest authoritiy on these subjects. He regards those fibres that unite the gray matter of the cerebrum with the large cerebral ganglia as the projection system of the first order.

Those fibres which pass between the cerebral ganglia and the gray matter around the ventricles, he considers the projection system of the second order. These are motor fibres passing through the crus cerebri and the pons into the white substance of the cord. The nerve fibres composing the roots of the

FIG. 136. Transverse section from a sulcus of the 3d frontal convolution of man. x 100. (Changed from Meynert). The medullary substance is not given in the drawing.

cerebral nerves he regards as the projection system of the third order.

GRAY MATTER.

Meynert classifies the gray matter under the following four divisions :

1. The cortex of the cerebral hemispheres.

2. The large cerebral ganglia
$\left\{ \begin{array}{l} \text{Corpora striata.} \\ \text{Corpora albicantia.} \\ \text{Optic thalami.} \\ \text{Corpora quadrigemina.} \end{array} \right.$

3. The gray substance of the medulla.
$\left\{ \begin{array}{l} \text{Rhomboidal fossa.} \\ \text{Aqueductus sylvii.} \\ \text{Tuber cinereum.} \\ \text{Gray matter lining ventricles.} \\ \text{Infundibula.} \end{array} \right.$

4. The cortex and central gray matter of the cerebellum.

Here in the gray matter is a fibrillar net-work corresponding to Gerlach's nerve net-work in the spinal cord. Multipolar ganglion cells are met with everywhere, varying in size and generally possessed of one unbranched process, the axis cylinder, which becomes a medullated nerve fibre. In the cortex of the cerebral hemispheres are the following five layers of Meynert :

1. Neuroglia and nerve net-work with small multipolar ganglion cells.

2. Small, pyramidal, closely crowded ganglion cells.

3. Numerous large ganglion cells, not crowded. These cells have (a), a branched process directed towards the surface, (b) branched lateral processes, (c) an axis-cylinder process in the centre of the basis processes.

4. Small irregular ganglion cells with few branched processes. He regards the first three layers as containing the motor cells and the last layer as connected with sensory nerves.

5. Branched, spindle-shaped ganglion cells parallel to

the surface. All of these cells have a nucleus and generally a nucleolus.

Meynert gives the following deviations from this rule ·

1. In the gray matter of the posterior portion of the
occipital lobe about the sulci hippocampi there are eight
layers. The prominent feature here is small multipolar cells
"the granular formation of Meynert."

2. In the cortex of the hippocampus major the small
cells of the fourth layer are wanting. The second and
third layers are the chief elements.

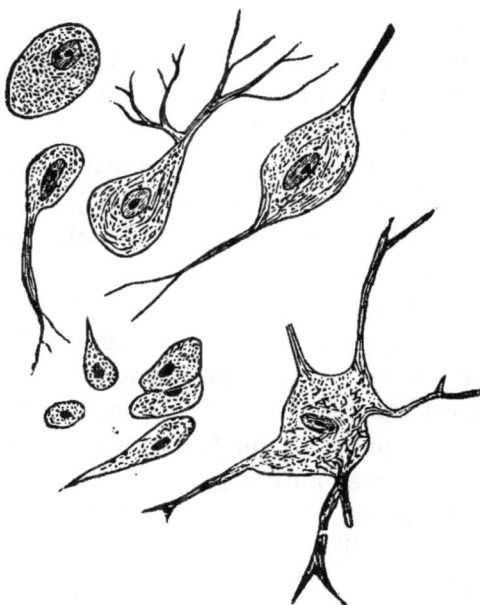

FIG. 137. Ganglia cells from cerebral convolutions. x 400.

3. In the walls of the fossa sylvii the fifth layer is most
prominent.

4. In the bulbus olfactorius there is a central cavity
lined with ciliated cells. The upper part is composed of
white matter, the lower of gray matter. This gray matter
consists of the following four layers from below upwards. A,

non-medullated nerve fibres which pass into the olfactory nerve. B, a layer of glomeruli; each glomerulus consisting of a convolution of an olfactory nerve fibre with many nucleated Deiter's cells. C, multipolar ganglion cells, spindle or pyramidal shaped. D, a net-work of fibrils with groups of nuclei.

In all these ganglia the cells are multipolar; spindle-shaped in the optic thalami and containing pigment in the corpora striata.

CELLS OF THE GRAY MATTER.

The ganglion cells of the gray matter are multipolar, and by their axis-cylinder processes give origin to the cerebral nerves. They are collected together in groups, each group giving origin to some particular nerve. The group of cells or "nucleus" of the optic nerve is a collection of multipolar cells of many sizes. The "nucleus" of the motory root of the fifth is situated in the anterior portion of the fossa rhomboidalis. They are large multipolar cells containing pigment. The cells are smaller in the sensory nucleus. In the "nucleus" of the facial, in the outer part of the superficial olivary body and on the floor of fourth ventricle, are very large multipolar cells. The "nucleus" of the acoustic nerve is in the fossa rhomboidalis near the surface. In the lateral anterior part of this fossa is the "nucleus" of the abduceus. Situated in this fossa also are the "nuclei" of the glosso-pharyngeal and vagus. The cells are spindle-shaped. The cortex of the cerebellum shows five layers.

1. A matrix of fibrillar nerve net-work and fine branced processes passing to the surface from the deeper layer of cells.

2. Large, spindle-shaped ganglion cells, Purkinje's cells. These cells have two processes. A branched process extending into the above layer and an unbranched or axis-cylinder process passing deeper.

3. A nuclear layer. This is between the second layer and the white substance. It is composed of minute fibrils with a great number of spherical nuclei. The corpora dentati

and olivary bodies are composed of a fine nerve net-work and slender multipolar ganglion cells. The brain is rich in blood-vessels.

METHODS OF EXAMINING.

Two methods are quite useful for examining fresh brain. Each of these methods has its advantages.

The first method will be found in full in the September number of the Monthly Microscopical Journal, vol. XVI, page 105, by Bevan Lewis. His method is briefly this: There are three stages of the process.

1. The preparatory stage, which consists in making as thin vertical sections as possible of the gray matter. The specimen is held in one hand between the thumb and fingers and with a sharp razor in the other, by a sweeping cut tolerably thin sections can be obtained. The upper surface of the knife should be deeply concave and kept flooded with alcohol. These sections are floated on a slide and a few drops of Müller's fluid placed over them. This is allowed to cover them completely and to make a pool around them for some seconds. The cover glass is then applied and by aid of a pencil or strongly mounted needle, steady gentle pressure is made on the centre of the cover until the nervous matter becomes a thin transparent film. The superfluous fluid is removed by rinsing in water and the slide is then transferred to alcohol. In about thirty or forty seconds the slide is removed from the dish of alcohol and while one edge of the cover-glass is steadied by the fingers, the blade of a penknife is gradually inserted beneath the opposite edge. The thin film will be left floating on the glass-slide or adhering to the cover-glass. The specimen is washed to free it from spirit by inclining the slide and allowing drops of water from a large camel's hair brush to flow over it.

2. The staining stage. A drop of a 1 per cent. solution of aniline black is placed on the film and as soon as the re-

quisite color has been acquired, the slide is transferred to a vessel containing water and gently lowered in it. By gently moving the water above it with a brush the superfluous dye floats away. Other staining agents, notably carmine, may be used.

3. The mounting stage. All fluid is drained off the specimen and it is placed under a bell jar with sulphuric acid. When perfectly dry add oil of cloves. This is removed when it has rendered the film transparent and dammar added and then the cover glass.

The Sankey method. Another excellent method was described in the April number of the Quarterly Microscopical Journal of 1876. Sections are cut as by the above method, only they may be as thick as the one-tenth of an inch. They are stained in a 7 p. c. solution of aniline blue-black. In three hours the staining is removed and water added until it washes away the excess of the staining. The specimens are floated on a clean slide, allowed to drain, and then exposed to the air in a dry place for twenty-four or forty-eight hours. At the end of this time the section will be firmly dried to the glass. It is now in a condition to have its upper surface planed off with a razor or other suitable instrument, taking care not to scrape the specimen away at any place. Dammar is added (oil of cloves not necessary) and the cover-glass applied permanently if an examination of the specimen is satisfactory. The nerve-cells usually show to good advantage by this method.

The brain may be hardened by processes recommended for the spinal cord. A 2 or 3 p. c. solution of bichromate of ammonia is a useful hardening agent. Small portions placed in Müller's fluid and then in alcohol are suitable for study.

CHAPTER XVIII.

Testicle and Ovary.

THE TESTICLE.

THE testes are small glandular bodies from one and one-half to two inches in length, one inch in breadth and one and one-fourth inches in their antero-posterior diameter.

The testes appear very early in fœtal life as two ovoid bodies situated at the inner borders of the Wolffian bodies. Two ducts are seen at their outer borders; the inner one, the duct of the Wolffian body, becomes the vas deferens of the male and disappears in the female; the outer one, the duct of Müller, becomes the fallopian tube in the female and disappears in the male.

The coverings of the testicle are described fully in the general works on anatomy. The minute structure of the gland substance only will occupy our attention.

When viewed with the naked eye the substance is of a reddish-yellow color, and is divided into a number of pyramidal lobes, the bases of which are directed toward the surface of the organ. Their number has been variously estimated at from 250 to 400. They vary in size according to the number of convoluted tubes they contain, for each lobe has from one to five or even more tubes—*tubuli seminiferi*—thrown into coils which are loosely held together, so that by careful dissection under water they can be disentangled to a considerable extent. It is estimated that 840 of these tubes are in each testicle and that each tube is on an average 30 inches in length and from $\frac{1}{200}$ to $\frac{1}{150}$ of an inch in diameter.

Each tube commences by several blind extremities or by anastomosing loops. Their walls consist of several layers of epithelial cells. The tubes from the different lobes sometimes anastomose with each other. Sometimes the several tubes of each lobe, together with tubes from the adjoining lobes unite together to form one canal. In this way about twenty canals are formed in each organ. At first these canals are quite tortuous but as they pass toward the posterior surface of the testes they become nearly straight— *vasa recta*—when they pass through the mediastinum forming a close net-work of tubes—the *rete testis*.

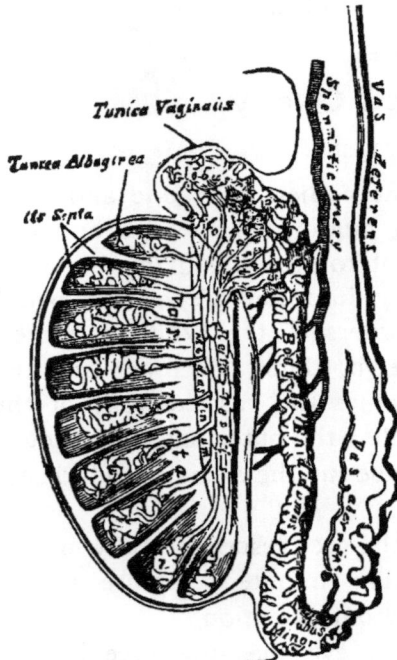

FIG. 138. Vertical section of the testicle to show the arrangements of the ducts. (Gray.)

The mediastinum is formed by a prolongation of the fibrous covering of the testis (*tunica albuginea*) into the posterior border of the gland and from this septum proceed the septa of fibrous tissue that divide the gland into its numerous lobes. Having traversed the substance of this fibrous septum the tubes leave the organ by from 12 to 20 canals known as the *vasa efferentia*. These tubes are about the $\frac{1}{50}$ of an inch in diameter and have walls of fibrous tissue and muscle cells.   At first these vessels are quite straight but as soon as they increase in size they become exceedingly convoluted.

The *epididymis* is the long narrow body lying along the outer edge of the posterior border of the testis.   It consists of an upper, enlarged extremity, the globus major, which is united to the testis by the vasa efferentia ; also a lower extremity, the tail or globus minor, and also a central portion or body.   After leaving the testicle the vasa efferentia form small cone-shaped masses—coni vasculosi—which together constitute the globus major of the epididymis.   These soon unite in one convoluted tube to form the body and globus minor of the epididymis.   When this tube is unravelled it measures not far from twenty feet in length ; at its commencement it is about $\frac{1}{70}$ of an inch in diameter ; it is reduced to $\frac{1}{90}$ of an inch before reaching the globus minor, however it soon increases in size again.   It is lined with ciliated columnar cells and its walls, thin at first, are thick and well provided with muscle cells toward its lower end.   The motion of the cilia is in an outward direction (Becker.)

After leaving the globus minor the duct turns upon itself to form the vas deferens.   This tube is about two feet in length and on an average $\frac{1}{8}$ of an inch in diameter.   It is very hard and cordy to the feel and it extends from the globus minor along the inner surface of the epididymis, behind the spermatic cord to the internal abdominal ring.   From here it passes into the pelvis, down the side to the base of the bladder, where it unites with the duct of the seminal vesicle of the corresponding side to form one of the ejaculatory ducts.   The vas deferens is lined with columnar epithelial cells, without cilia.

The seminal-vesicles—*vesiculæ seminales*—are composed of convoluted tubes, which, when unravelled, are three or four feet in length, and lined with short prismatic epithelial cells.

The *vas aberrans* of Haller is a small elongated mass, composed of a single convoluted tube, which is given off from the lower part of the epididymis as a diverticulum of the canal forming that body.   When unravelled the tube varies from two to fourteen inches in length.   It represents the remains of one of the tubes of one of the Wolffian bodies which still remains attached to the excretory duct.   In the commencing portion of the vas deferens is the organ of Girardès.   It is known also by the name of parepididymis ; it is made up entirely of canals.

The seminal tubes in the adult testis are lined with several layers of epithelial cells, called the seminal cells.

Those cells situated next to the walls of the tube are known as the outer cells and those situated near the lumen are the inner seminal cells.

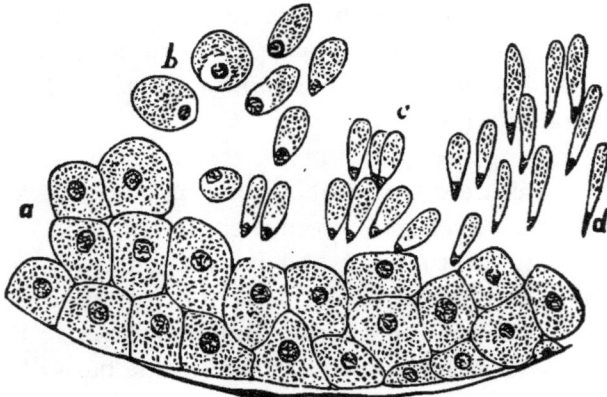

FIG. 139.   A portion of the wall of a seminal tubule of the testicle of the dog. a, seminal cells.   b, spermatoblasts.   c, earliest stage in the formation of the spermatozoa.   d, spermatozoa more fully developed.   (Klein and Smith.)   x 450.

Klein notices two kinds of cells in the outer layer : 1, those having an oval transparent nucleus limited by a membrane and provided with one, two or three nucleoli ; 2, those with a spherical nucleus, the cells smaller than the former, and the nucleus not limited by a membrane.

The inner seminal cells contain in their matrix a convolution of rods, twisted in various directions and anastomosing with one another. The nuclei of the cells nearest the lumen of the tube very frequently undergo division, the cells themselves dividing afterwards. Not unfrequently multinuclear cells are seen with from six to ten nuclei. Each nucleus is spherical in shape and is not limited by a membrane.

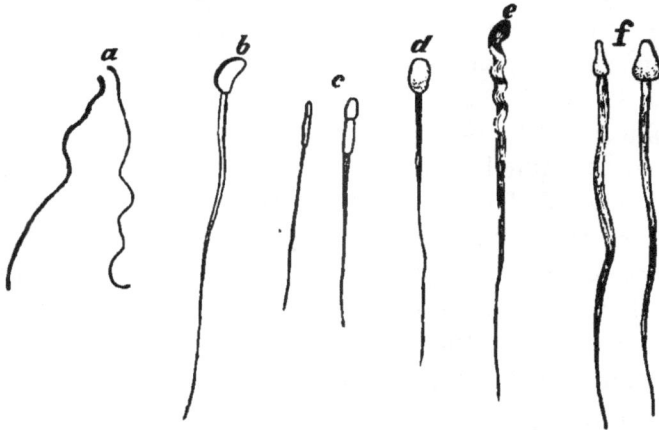

FIG. 140. Spermatozoa. a, of the blaps mortisago. b, of the house-mouse. c, of the bat. d, of the sheep. e, of the Raja batis. f, of man.

As these cells become developed into the spermatozoa, they have been named by Sertoli, the spermatoblasts. Studying one of these cells with a spherical nucleus, it is noticed, first, that the nucleus becomes limited by a membrane and is placed near the edge of the cell. The nucleus looses its reticulated appearance and its substance becomes collected in one part so that a clear space, a clear tube, is left. The nucleus becomes gradually more disc-shaped, and the cell substance is drawn out into a club-shaped granular body which is separated from the nucleus by the clear space mentioned above (Klein.) By other gradual changes the epithelial cell finally becomes a fully developed spermatozöon, which has a well defined oval head, behind which is the middle-piece of Schweigger-Seidel, attached to which is the tail ending in a very fine pointed extremity.

The nucleus of the inner epithelial cell has become the head of the spermatozöon, the granular body has become the tail, and an outgrowth of the nucleus has become the middle-piece.   From the researches of H. Gibbes, [Quarterly Journal of Microscopical Science, Oct., 1879] we learn that each spermatozöon of many of the vertebrates consists of [1] a long, pointed head, [2] an elliptical structure joining the head, [3] a filiform body, [4] a fine filament much longer than the body and connected to it by [5] a homogeneous membrane. When living this filament is in constant motion and has a continuous waving from side to side.   He is "confident that the substance of which the head is composed shows a different chemical reaction to the rest of the organism."   The waving filament was seen in the spermatozoa of the following mammals: The horse, dog, bull, cat, rabbit, guinea-pig, and man.

He draws the following conclusions :

1.   That the head of the spermatozöon is enclosed in a sheath which is a continuation of the membrane which surrounds the filament and connects it to the body.

2.   That the substance of the head is quite distinct in its structure from the other parts and that it is readily acted upon by alkalies ; these reagents having no effect on the other parts except the membranous sheath.

3.   That the motive power lies in a great measure in the filament and the membrane attaching it to the body.

THE OVARY.

The tissues of the hilum consist of connective-tissue, blood-vessels, muscle cells, etc.

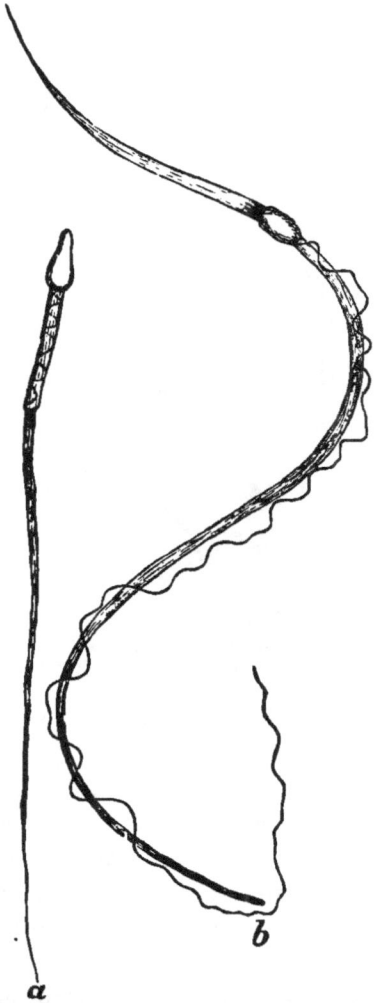

Covering the free surface of the ovary is a layer of short, columnar, nucleated epithelial cells, the "germinal layer" of Waldeyer. In the cortical portion of the ovary there is a layer of tissue free from ova , this has been named by Henle the tunica albuginea. Spindle-shaped cells are met with as well as branched cells. Besides these cells there are groups of nucleated, polyhedral cells ; Balfour regards these cells as the remains of the epithelial cells of the Wolffian bodies, as do also His, Waldeyer and others. The Graafian follicles are of many sizes and shapes. The smaller ones are situated near the surface, just beneath the albuginea ; while the larger ones are in the deep seated portions of the organ. The small follicles are so closely and densely arranged just under the albuginea that they constitute a distinct layer, the "cortical layer" of Schrönn. Around the larger follicles the

FIG. 141. Two fully developed spermatozoa. a, of the horse. b, of the triton cristatut.
(H. Gibbes.)

spindle-shaped cells are arranged concentrically, forming an investment to the follicle, the "tunica fibrosa" of Henle. Each follicle is surrounded by a membrana propria, in which are seen a few nuclei. Just beneath this layer is one of epithelial cells, the membrana granulosa, and within these cells is the ovum. The membrana granulosa varie from a single layer of flat cells in the smaller follicles to columnar shaped cells of one, two or more layers in the larger ones.

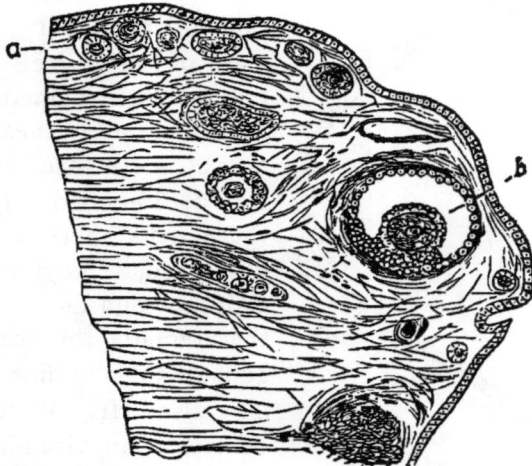

FIG. 142. Portion of a section of a cat's ovary. a, layer of small ovigerms. b, a follicle farther advanced. The ovum is covered with the cells of the discus proligerus. x 35.

The ovum of the fully developed follicle lies in a mass of cells which project from the membrana granulosa like a mound. This projection is called the discus proligerus. In the smaller follicles the ovum is surrounded by a mass of cells which are everywhere in close contact with the membrana propria, but soon a fluid appears between some of the cells, causing little cavities which increase in number and size until these several inter-cellular cavities become confluent and the ovum at last becomes separated from the membrana propria, except where it remains connected by the intervening discus proligerus. This fluid is called the liquor folliculi. In this fluid are seen occasionally a few small, more or less vacuolated cells. They once belonged to the follicular epithelium, but are now undergoing retrogressive changes [Klein] and will soon disappear.

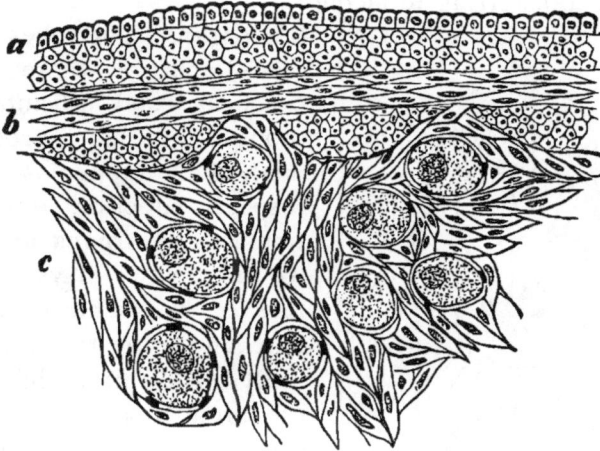

FIG. 143. Vertical section through the ovary of a half-grown cat. a, germinal epi-thelium of the surface. b, albuginea. c, Graafian follicles. x 350. (Klein & Smith.)

The ovum is the nucleated cell embedded in the discus proligerus. Its nucleus is known as the " germinal vesicle " and its nucleolus or nucleoli as the " germinal spot or spots."

The protoplasm immediately surrounding the nucleus is more transparent than that at the periphery of the ovum. Surrounding the ovum is a clear, transparent ring, the zona-pellucida, which, according to Waldeyer and others, often shows in the mature ovum a vertical striation caused by the continuation as fine threads of the epithelial cells surrounding this zone.

" Many follicles arrive at the stage of ripeness before puberty is reached and are subject to a process of degeneration. But this process involves also follicles of earlier stages and even the smallest follicles." [Klein and Smith, Atlas of Histology, p. 290, 1880.]

METHODS OF EXAMINING.

The testis. The familiar process of hardening in Müller's fluid and transferring to alcohol is a very satisfactory one for this organ. The tubes can be teased apart in these preparations to good advantage, although for this purpose the method of Sappey is in some respects superior, small portions of the organs are placed for one or two days in a mixture of hydrochloric acid, one part, and water two parts.

By the aid of a fine syringe the organ may be injected at several points with a 1 per cent. solution of osmic acid. The specimen is then placed in alcohol, when after a few days it is sufficiently hardened to admit of thin sections being made in various directions. This process is highly recommended by Mihalkovics.

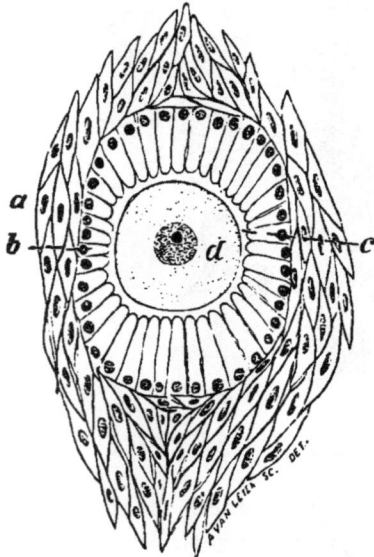

FIG. 144. Graafian follicle. a, spindle-cells of stroma. b, membrana propria et granulosa. c, zona pellucida. d, ovum with its germinal vesicle and germinal spot. x 350. (Klein and Smith.)

The Ovary. Beautiful sections of the ovary of the cat are in the author's possession, made from specimens hardened in Müller's fluid and alcohol, stained with carmine, cleared in oil of cloves and mounted in dammar.

From the fresh ovary of the dog or rabbit, ova may be obtained as follows : While the ovary is held firmly in the hand, the most prominent Graafian follicle is pricked and its contents received on a glass slide in a drop of normal fluid. If present the ovum is easily recognized when it should be examined, uncovered and with a low power. Afterwards a higher power may be substituted, covering the specimen, but inserting beneath the cover a piece of paper to avoid pressure.

CHAPTER XIX.

The Tongue, Skin, Lining of the Nasal Cavity and the Ear.

THE mucous membrane lining the mouth is covered with a thick layer of stratified epithelial cells ; and among these cells are a few that are connected together by fine filaments, although they are separated from each other by the ordinary cement-substance. When the parts are inflamed, the intercellular substance is increased and the connecting filaments show to much better advantage. These are the "prickle-cells" of M. Schultze. In the mucous membrane are numerous large mucous glands. The duct of each gland is a single layer of endothelial cells and it is a direct continuation of the basement membrane, which here, as in other parts of the alimentary canal, is between the layer of epithelial cells and the mucosa. Lining the duct is a layer of nucleated columnar cells. Upon reaching the submucous tissue the duct divides into a number of branches. These branches are convoluted, and they have one or more short lateral branches. The body of the gland is lined with cells identical with those lining the mucous salivary glands. When the gland is inactive these cells are transparent and are filled with mucigen which is changed into mucin during secretion.

The tongue is essentially a muscular organ, covered with mucous membrane. On its dorsal surface are three kinds of papillæ; the circumvallate, fungiform, and filiform. The first are seen with the unaided eye at the base of the tongue, arranged in the form of a V. The second are distributed over the surface of the tongue and are visible also to the unaided eye. The third are the most numerous as well as the smallest. They are not especially concerned in the sense of taste. In the circumvallate and in many of the fungiform papillæ there are peculiar bud or flask shaped organs—the organs of taste— the gustatory buds—the taste buds of Schwalbe. They are about $\frac{1}{300}$ of an inch high in man, and they exist in large num-

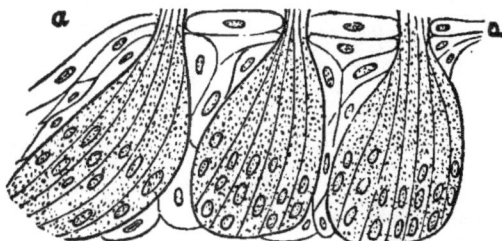

FIG. 145. Taste buds, a, epithelial surface. (Klein and Smith.) x450.

bers. They are covered externally with one, usually two or more, layers of long, tapering, flat cells, which are in close contact with each other. They surround the opening of the organ above and "stand like the staves of a barrel." These peripheral cells surround the central or taste cells. The central cells are slender, nucleated and spindle-shaped. The nucleus causes an enlargement of the cell at that point. The cells terminate above in a rod shaped extremity at the opening of the organ; below they terminate in a slender often bifurcated extremity which, it is supposed, passes as a nerve axis or axis fibrilla into the gustatory nerve. Fine nerve fibrils are certainly connected with these taste cells. (Englemann, Hömgschmied.)

Situated near the taste buds are serous glands, embedded in the mucous membrane. The bodies of these glands are like the parotid in structure. There are no mucous cells found in them. In all probability they secrete a watery substance which is poured over the parts containing the taste buds and thus they assist in the distribution of the substance to be tasted.

The nerves of the papillæ contain end bulbs (Krause), and the tactile corpuscles of Meissner can be demonstrated, (Geber).

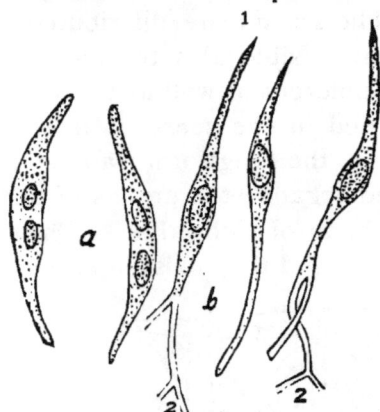

FIG. 146. Cells from taste buds. a, cover cells. b, cells from central part. 1, the fine extremity which projects at the orifice of the bud 2, the deep extremity which becomes continuous with fine nerve fibrils. (After Englemann and Klein.)

MUCOUS MEMBRANE OF THE NOSE.

Only a portion of this membrane (Schneiderian) is of especial interest in this connection. The portion which contains the terminations of the olfactory nerve is distributed over the upper portion of the septum, and the upper and middle turbinated bones. Nonciliated epithelial cells cover this highly vascular membrane in these regions, although the ciliated variety abounds in the other parts. Glands are scattered freely through the membrane.

The yellowish appearance of this membrane is due to the presence of yellowish granules in the long columnar cells covering its surface. These cells send long processes downwards, which usually give off numerous branches which anastomose with other branches (Martin) from neighboring cells, to make a continuous net-work just beneath this layer. Between these cells are nucleated, spindle-shaped ones—the olfactory cells of Schultze. One end of these cells terminates in a fine process on a level with the surface of the columnar cells between which they lie; the other end terminates below in a fine filament which may be directly connected with a fine axis fibrilla from one of the non-medullated fibres of the olfactory nerve.

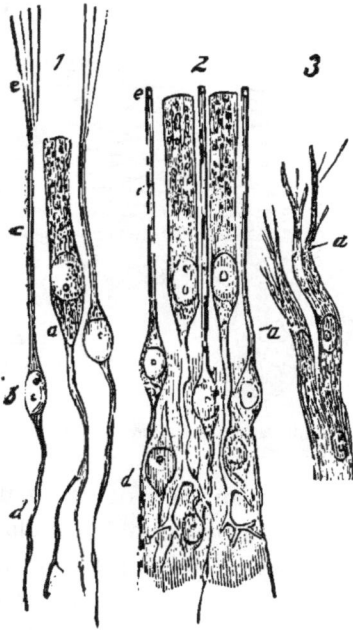

THE SKIN.

The importance of this complex structure of the body is very evident when we consider the variety of its functions and the extent of its surface. It forms a protective covering for the body; it is an organ of tactile sensibility, it preserves the external forms of the muscles; it is an organ of excretion, and it aids in maintaining the normal temperature of the body. In the case of an average size man its surface is equal to sixteen square feet, and of an average size woman, twelve square feet.

FIG. 147. 1, Cells of the regio olfactoria of the frog. a, an epithelial cell, terminating below in a ramified process. b, olfactory cells with the descending filament. d, the peripheral rod, c, and the long vibratile ciliæ, e. 2, cells, from the same region of man. The references the same, only short projections, e, occur (as artefacts) on the rods. 3, fibres o the olfactory nerve from the dog. At a, dividing into fibrillæ. (Quain.)

The skin may be divided into two principle layers: 1, the true skin or corium, and 2, the cuticle or epidermis.

The epidermis is divided into the following four layers:

1. Rete-Malpighii or rete mucosum.
2. Granular layer (of Langerhaus.)
3. Stratum lucidum (Schrön.)
4. Stratum corneum.

The deep cells of the first layer are columnal and provided with nuclei. The more superficial cells are polyhedral, also nucleated.   The lower surface of this layer is uneven and presents papillæ and corresponding depressions, which adapt

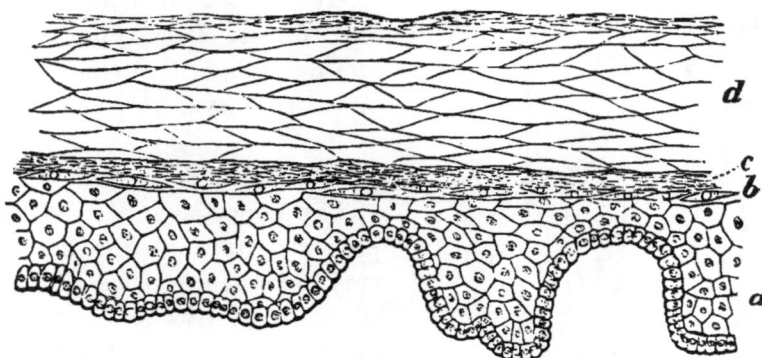

FIG. 148.   Epidermis   a, rete-mucosum.   b, granular layer.   c, stratum lucidum. d, stratum corneum.   (Smith and Klein.)   x350.

themselves to the corresponding uneven surface of the true skin.   The layer of Langerhaus is composed of granular, spindle-shaped cells which take hæmatoxylin staining to an in-

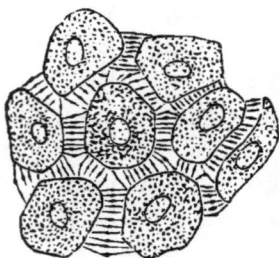

tense degree.   The third layer is a translucent one, occasionally showing in it a few closely packed cells.

The fourth layer, the most superficial, is made up of many layers of horny epithelial cells, without nuclei, as a rule.

In colored skins the pigment is deposited around the nuclei of the deeper cells of the rete-Malpighii.

FIG. 149.   Polyhedral cells of the rete-Malpighii connected with each other by fine filaments.   Prickle cells of Schultze.   (after Klein and Smith.)   x550.

(Klein.)   In this layer are found the prickle-cells of Schultze. The corium, derma, or cutis vera, is protected above by the epidermis and is attached to the parts beneath it by a loose areolar tissue.   Fibrous tissue forms the bases of the true skin.   Distributed through it are blood-vessels, nerves, glands, hair bulbs, etc.   The upper part of this layer is divided into papillæ which are received in the depressions between the descending papillæ of the rete-Malpighii.   In those parts where the sense of touch is most delicate there are found the largest number of papillæ supplied with terminal nerve fibres.   These are the tactile corpuscle, (see figure 128), found in great abundance in the skin covering the palms and soles of the hands and feet; less abundant in the skin of the palmar surface of the forearm and the nipple,   Pacinian bodies are found in the subcutaneous tissue and "end bulbs"on the glans penis and glans clitoridis, and occasionally on the borders of the lips.   Aside from these special nerve terminations there is a superficial plexus of delicate nerve fibrils just beneath the rete-mucosum.   Many of the cutaneous nerves are distributed to the hair follicles.

The vascular papillæ are supplied with either a single branch which terminates in a loop, or with several branches which anastomose and then return to the veins.

F G. 150.   A human sudoriferous gland.   (After Klein.)

The lower part of the corium is composed largely of white fibrous tissue in which are seen a few muscular and elastic fibres.   The hair sacs have been described already, (see hair.) Situated as a rule in the subcutaneous cellular tissue are the sudoriferous or sweat glands.   They consist of a fine tube coiled up after the manner of our figure.   The tube is lined throughout with epithelial cells.   In the walls of the tubes are smooth muscular elements. The duct is twisted upon itself and opens on the surface of the skin with a funnel shaped dilation.

The sebaceous glands or submucous follicles are situated

in the corium generally near the hairs.   The sacs composing
the gland are lined with epithelial cells as are their ducts.
They open into the hair follicles discharging a thick fatty substance.   They frequently become enlarged through a closure
of their ducts, as seen on the face and alæ of the nose.

THE EAR.

The minute structure of the ear is so very complicated
that it would carry us beyond the limits of this work to give
anything like a full description of its several parts.   The following abridged account will enable the student to partially
understand the terminations of the auditory nerve.   The
essential part of the ear is situated in the petrous portion of
the temporal bone, and is known as the internal ear or labyrinth.

The osseous labyrinth consists of three parts ; the vestibule, the semi-circular canals, and the cochlea.   Within this
labyrinth, but separated from it by a clear fluid, the perilymph,
are membranous structures in which are found the terminal
fibres of the auditory nerve.   This membranous labyrinth encloses a fluid called the endolymph.

The vestibule communicates externally with the tympanic
cavity, internally with the internal auditory meatus, anteriorly
with the cochlea, and posteriorly with the semi-circular canals.
Five openings unite the vestibule with the semi-circular canals.
While one aperture—apertura scalæ vestibuli—unites it with
the cochlea.

The semi-circular canals are three in number.   They are
of unequal length and they measure $\frac{1}{20}$ of an inch in diameter.
At one end of each canal is an enlargement called the ampulla
which is more than twice as large as the tube.   The cochlea is
about one-fourth of an inch in length, and in breadth toward
the base it is the same.   It bears some resemblance to a common snail shell.   Its conical-shaped central axis is called the

modiolus, which extends from the base to the apex.   Its base
is perforated with numerous orifices through which pass deli-

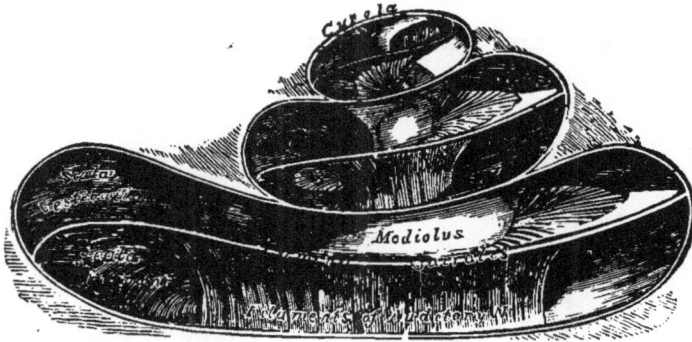

FIG. 151.   The cochlea laid open.   (Enlarged.)   (Gray.)

cate nerve fibrils from branches of the auditory.   One of the
orifices larger than the rest is called the tubulus centralis mo-
dioli.   It extends the whole length of the modiolus and trans-
mits a nerve and artery.

Surrounding the modiolus in a spiral manner for two
turns and a half is the spiral canal—laminis spiralis ossea.   It
is about one and a half inches in length and one-tenth of an
inch in width at its commencement.

It gradually diminishes in size on its way to the apex
where it terminates in a closed extremity to form the apex of
the cochlea, called the cupola.   This canal is partly divided into
two passages or scalæ, by a thin flat osseous, membranous
plate, which winds around the modiolus and projects into the ·
spiral tube.   These scalæ are known as the scala tympani and
scala vestibuli.

Within the osseous labyrinth is the membranous one,
smaller than the former, and separated from it by the perilymph.   It has a general resemblance in form to the vestibule
and semi-circular canals.   It is loosely united to the lining

membrane of these parts by fibrous bands. The membranous labyrinth forms a closed sac containing the endolymph. In the vestibule the membranous labyrinth consists of two parts, the utricle and the saccule. Attached to the walls of these sacs are crystals of carbonate of lime, otoliths or otoconia.

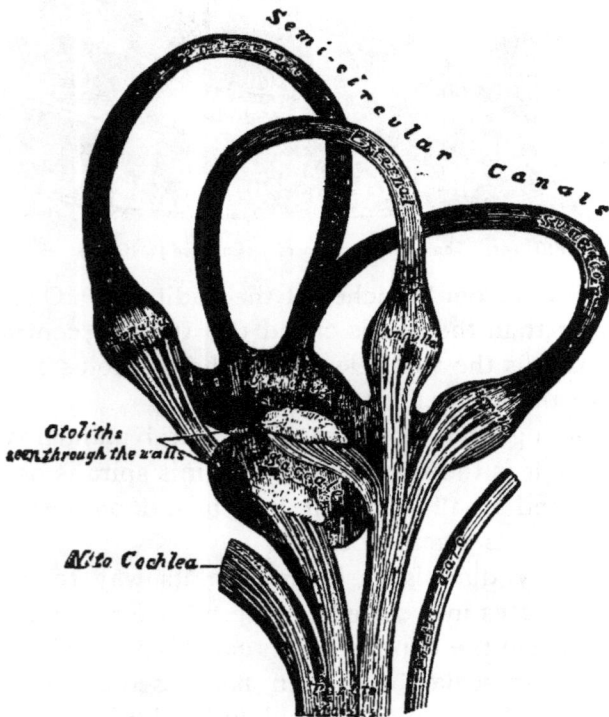

FIG. 152. The membranous labyrinth detached. (Enlarged.) (Gray.)

Projecting into the cavity of each of the ampullæ is a rounded eminence, caused by a thickening of the tunica propria. Over this eminence the epithelial lining of the ampullæ consists of cells of the columnar shape. Between these cells

FIG. 153. Otoliths.

are spindle-shaped ones which have hair-like processes that project into the endolymph.

Branches of the auditory nerve reach the tunica propria of the ampullæ, when the axis cylinders divide into their primitive fibrillæ. These fibrils form a network just beneath the attached ends of the epithelial cells. A single nerve axis fibril now passes through the long axis of each of the spindle-shaped cells and terminates in the free hair-like processes (Rudinger). Relzius gives a different view of these termi-nal fibres. He regards the hair-like processes as belonging to the colum-nar shaped cells, and that the narrow end of each of these cells becomes continuous with a nerve fibril. The membranous labyrinth of the cochlea consists of a tube lined with epitheli-al cells and enclosing endolymph. This tube is divided into three parts by two membranes; one of these membranes is formed by the prolon-gation of the lamina spiralis to the outer wall of the cochlea and is call-ed the basilar membrane. Stretch-ing from this membrane obliquely across to the outer cochlear wall is the delicate membrane of

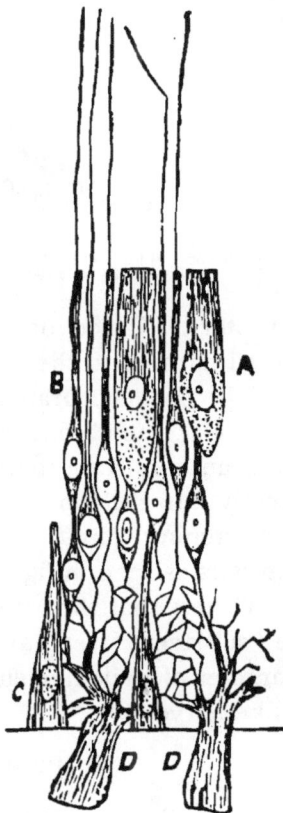

FIG. 154. Diagram of the auditory epithelium and the mode of termination of the nerves of the ampullæ. a, columnar epithelium. b, spindle-shaped cells each sup-porting a hair. c, basal supporting cells. d, two nerve fibrils. (Quain after Schultze.)

Reissner, by its direction enclosing a triangular space, ductus cochlearis.

The epithelium covering the basilar membrane contains the so-called organ of Corti. This forms a conical elevation and is hollow in its interior. In the centre of this organ are two sets of rod-like bodies, rods of Corti, which are so arranged as to enclose a space, filled however with endolymph in the natural condition.

On the inner side of each inner rod is a long cell covered at its free border with short hairs. This is the inner hair cell

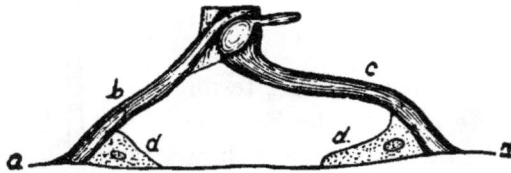

FIG. 155. A pair of rods of Corti, side view, from the rabbit. a, basilar membrane. b, inner rod. c, outer rod. d, nucleated protoplasmic masses. (Quain.)

of Deiters. Three or four rows of outer hair cells are arranged parallel to the "outer rods.

The cochlear branches of the auditory nerve pass through the basilar membrane and are continued as axis-cylinders. Some authors state that a single primitive axial fibril passes directly into the inner hair cells. Others describe these fibrils as stretching across the canal and terminating in the outer hair-cells as seen in figure 156.

Others that the termination is not in the hair cells, but in the "subjacent irregular protoplasmic cells which both in the character of their nucleus and in other particulars are not very unlike nerve cells."

METHODS OF EXAMINING.

The Tongue. The tongue is best studied in hardened

injected specimens.   The tongue of the rabbit can be injected through the arteries with Prussian blue, hardened in a two per cent. solution of potassic bichromate or in a weak solution of osmic acid and the thin sections tinged with hæmatoxylin.

The sections should be cut vertically to the mucous membrane, and mounted in glycerine or dammar.

The sections show to the best advantage if a double staining be practiced.   To accomplish this the sections are placed first in a saturated solution of picric acid for about twenty minutes, then they are placed in the hæmatoxylin.

To study the branched muscle fibres small pieces of the tongue are boiled in water a short time, when the fibres can be teased apart readily.

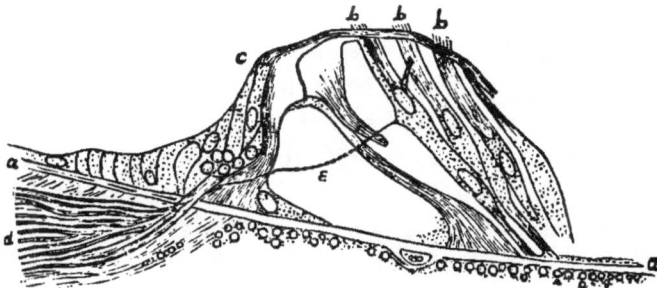

FIG. 156.   Organ of Corti, (dog).   a, basilar membrane.   b, outer hair cells.   c, inner hair cells.   d, nerve,   e, nerve fibril.   (After Waldeyer.)

The taste-buds are studied in specimens that have been immersed for several days in a two per cent. solution of chromic acid, to which has been added an equal volume of glycerine.   These preparations are carefully picked under the microscope.   Specimens that have been in Müller's fluid for 15 or 20 days are in good condition to be picked.

To study the nerve terminations chloride of gold (.5 p. c.), chromic acid (.02 p. c.), and osmic acid (1 p. c.) preparations are to be preferred.

The Skin.   The skin over the palms of the fingers is injected from some large vessel in the hand or fore-arm; and vertical sections are made either from the fresh specimens with the freezing microtome, or, better still, from specimens hardened in a mixture of chromic acid and alcohol [chromic acid ½ per cent. solution and alcohol an equal volume].   After remaining in this mixture for two weeks the pieces are transferred to alcohol for two or three days, when they are ready for making sections.   Double staining with the picric acid and hæmatoxylin, tinging with carmine, etc., will prepare the specimens for mounting in glycerine or dammar.

The Ear.   The following is Dr. Pritchard's method for preparing the cochlea. (Abbreviated from Beale. Mic. in Medicine, 4th ed. pp. 438, 439.)   "The cochlea should be as fresh as possible.   As much as possible of the bone around it should be removed.   The membrane should be hardened by maceration in a ¼ per cent. solution of chromic acid.   From three weeks to a month will suffice in this mixture.   The bone is softened by adding from a ½ to a 1 per cent. solution of nitric acid to the chromic acid solution during the last few days of maceration.

A large quantity of the solution should be used and it should be changed every four or five days.

When the bone is quite soft it must be prepared for section in the following way :   Make a small conical bag of paper, fill it with strong gum-water, and put in this softened cochlea (taken directly from the chromic acid solution,) leave it in the gum-water to soak for a few hours and then place the whole (paper bag and all) in alcohol.   At the end of twenty-four hours remove the bag from the spirit, and pick away the paper and gum, which will then be quite white and hard. Should the cochlea now be found not sufficiently firm, it may be steeped in absolute alcohol for a minute or two and then it will be ready for section.

For holding the cochlea in the microtome the following mixture is most convenient :

Paraffine, - - - - - - 5 parts.
Spermaceti, - - - - - - 2 parts.
Lard, - - - - - - - 1 part.

Melt in a water bath. Before mounting dissolve away the gum with water and put up in glycerine or in a saturated solution of acetate of potash made with camphor water."

CHAPTER XX.

The Eye.

THE posterior five-sixths of the eyeball presents a sphere on which rests the anterior one-sixth as a segment of a smaller sphere. The posterior sphere is opaque and is covered with the sclerotic; the anterior sphere is transparent and covered with the cornea.

The sclerotic is composed of dense fibrous tissue and varies in thickness in its different parts. At its anterior margin where it joins the cornea it is about $\frac{1}{30}$ of an inch thick ; over the greatest convexity of the eyeball $\frac{1}{60}$ of an inch ; where it is pierced by the optic nerve $\frac{1}{25}$ of an inch. It has few blood-vessels. On the inner surface of the sclerotic and uniting it with the choroid is a layer of connective tissue known as the "membrana fusca." It is continuous posteriorly with the sheath of the optic nerve.

The cornea is about $\frac{1}{25}$ of an inch thick where it joins the sclerotic, but thinner in its central portion. Covering it on its free surface is a layer of pavement stratified epithelium. The superficial cells of this layer are flattened ; beneath these cells are smaller ones nearly round ; while beneath these there are cylindrical ones arranged perpendicular to the surface. These cells are all nucleated. Beneath this epithelial covering is a thin transparent membrane "anterior elastic lamina,"—Bowman's membrane. It is a continuation of the conjunctiva and is about $\frac{1}{2000}$ of an inch in thickness. Covering the posterior surface of the cornea is a single layer of polygonal nucleated cells. Anterior to this layer is a firm, elastic, transparent membrane, "membrane of Descemet or Demour." It varies in thickness from $\frac{1}{3000}$ to $\frac{1}{2000}$ of an inch. The *ligamentum iridis pectinatum* is composed of processes from this membrane which have passed to the anterior surface of the iris. These processes are covered with epithelial cells which are continued directly from those on the posterior surface of the cornea.

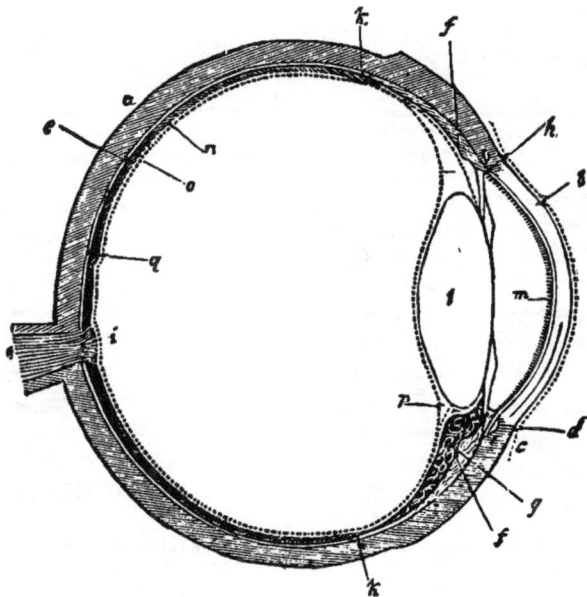

FIG. 157. Transverse section of tne eye. a, sclerotica. b, cornea. c, conjunctiva. d, circulus venosus iridis. e, choroid, with the pigment layer of the retina. f, ciliary muscle. g, ciliary process. h, iris. i, optic nerve. j, colliculus opticus. k, ora serrata retinæ. l, crystalline lens. m, tunica Descemetii. n, membrana limitans interna of the retina. o, membrana hyaloidea. p, canalis Petiti. q, macula lutea. (Selected.)

While these cells cover the processes they do not extend between them, so that the anterior chamber is prolonged into the spaces between these processes. Other processes are given off to the sclerotic and choroid. At the margin of the sclerotic, close to its junction with the cornea is a circular opening, the canal of Schlemm. Between the two transparent layers is the corneal tissue proper. This consists of pale bundles of fibres, which cross each other in every direction. The bundles of fibres may be divided into the most minute fibrillæ. They become directly continuous with the fibres of the sclerotic and they cannot be separated from each other by maceration.

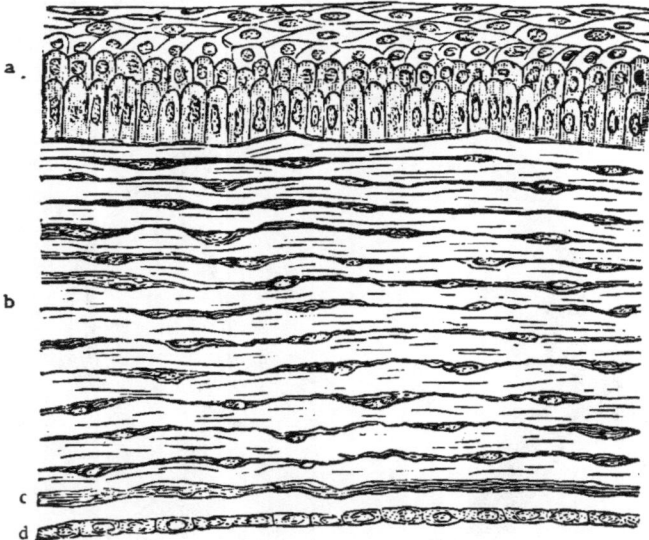

FIG. 158. Vertical section of cornea of rabbit, hardened in chromic acid. a, anterior layer of pavement epithelium. b, Substantia propria of the cornea, consisting of connective tissue fibres in more or less parallel bundles, between which are the cornea corpuscles. These, in vertical sections, appear spindle-shaped. c, the posterior lamina elastica, or Descemet's membrane, and the endothelium of polyhedral cells, d, which covers it. (Burdon-Sanderson.)

These bundles of fibres are separated from each other by a ground, or interstitial substance. It is in these parts that the cell spaces are found. The spaces are flattened, stellate, and freely communicate with each other by their processes. When viewed in a vertical section they appear fusiform, but in a horizontal section they are flattened, irregularly stellate spaces. Within these spaces are the corneal corpuscles ; they conform quite generally to the shape of the spaces, but do not completely fill them, thus leaving room for the lymph and the wandering cells. The cell substance is a clear protoplasm after the manner of that in the lacunæ and canaliculi of bones, with a nucleus surrounded by granular matter. This proto-plasm entends out into the processes of the cell spaces, anastomosing with like processes from neighboring cells.

The cell is capable of withdrawing and protruding these processes, the movements are slow, and are not to be taken

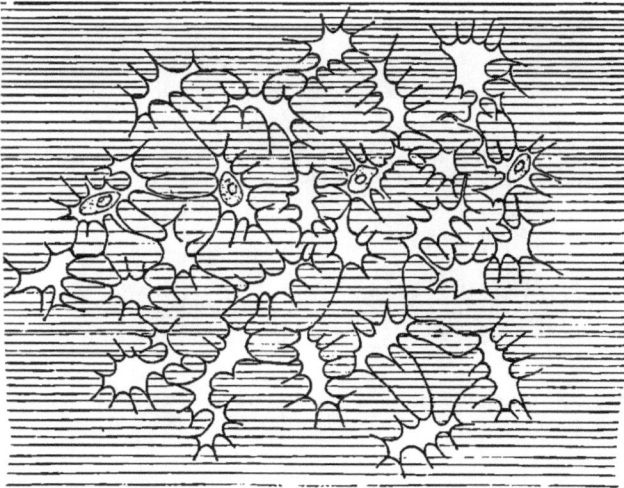

FIG. 159. Corneal corpuscles of human cornea, silver staining. x 500.

for the rapid amœboid movements of the leucocytes. By the puncturing method the system of cell spaces, can be fully injected in the cornea and continued into the lymphatics of the sclerotic. Bowman's membrane is regarded by many as a part of the corneal tissue proper, but without corpuscles. The healthy cornea is not provided with blood-vessels, for these terminate in loops at the circumference.

The choroid is situated beneath the sclerotic and is attached to it by the membrana fusca.   It is easily recognized
by its dark color.   It is about $\frac{1}{25}$ of an inch thick near its anterior termination which is very near the cornea ; here it terminates in a series of folds or plaits known as the ciliary processes.   At its middle it is only $\frac{1}{150}$ of an inch thick, but
thicker again at its posterior.   Its outer part consists of arteries and veins, arranged in a peculiar manner, described
and illustrated in the standard works on anatomy.   The veins
are external to the arteries and are arranged in curves, vasa
vorticosa.

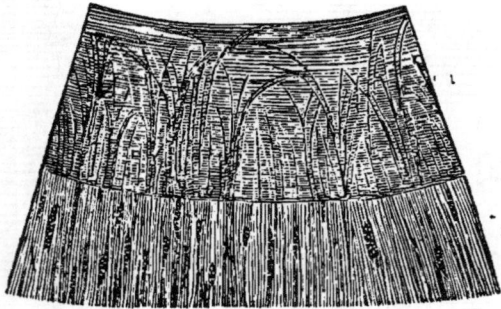

FIG. 160. Surface of the human iris, a, the sphincter.   b, the dilator of
the pupil.   (Iwanoff.)

Between the vessels are pigment cells with numerous fine
branches.   The outer part of this layer is surrounded by a
membrane—lamina supra-choroidea.   Between this membrane
and the membrana fusca is a lymph space, which communicates with the space existing between the eyeball and the thin
membrane surrounding it—the vaginalis oculi,—which is lined
with a single layer of cells, affording a smooth surface for the
eyeball to move against ; externally it separates the eyeball
from the fat of the orbital cavity.   The inner part of the

choroid consists of a layer of capillary vessels, the tunica Ruyschiana. Internal to this is the membrane of Bruch, a thin transparent membrane, exhibiting most minute folds or plaits at its anterior margin. The ciliary processes are folds or plaits of the choroid at its anterior extremity, projecting internally. There are from fifty to seventy-five of them, and the largest are about $\frac{1}{10}$ of an inch in length. Between these folds are smaller ones, formed by folds or plaits of the hyaloid membrane—the zone of Zinn

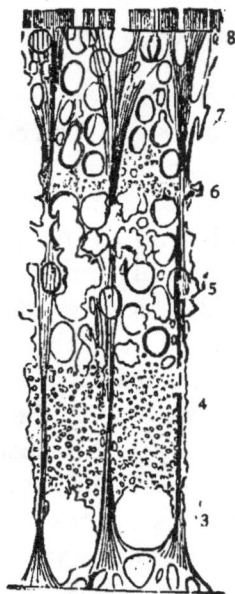

FIG. 161. Diagram of the connective substance of the retina. (Klein.)

The ciliary muscle surrounds the anterior margin of the choroid, arising by a thin tendon from the inner side of the boundary line between the cornea and sclerotic. The fibres radiate posteriorly and are inserted in the choroid, just outside of the ciliary processes. Some of the fibres pursue a circular course around the attachment of the iris ; this is the ring muscle of H. Müller. It is much developed in hypermetropic eyes, being atrophied or absent in myopic eyes [Iwanoff.] The muscle is about $\frac{1}{8}$ of an inch wide and $\frac{1}{50}$ of an inch thick at its thickest part.

The iris is the circular membrane just anterior to the crystalline lens. It gives the particular color to the eye, and with its perforated centre corresponds to the diaphragm of our optical instruments. It is about $\frac{1}{2}$ an inch across and is perforated a little to the nasal side of the centre. Although constantly varying in size, the pupil is ordinarily from $\frac{1}{8}$ to $\frac{1}{6}$ of an inch across. At its periphery it is connected with the choroid, cornea and ciliary muscle. The pupil is closed during a certain portion of fœtal life by a membrane called the pupillary membrane. It is not seen at an early period, but becomes most distinct at the sixth month, commences to break away at the seventh, and at birth no traces of it can be seen as a rule. The anterior surface of the iris is covered with a continuation of the epithelial cells that formed the posterior layer of the cornea. Just beneath this layer is a layer of irregular pigment cells. Covering the posterior surface is a layer of pigment cells also, a continuation of those covering the retina and lining the ciliary processes. The cells are small, rich in pigment, and arranged in several layers. The color of black, brown, and gray eyes is caused by pigment cells in the substance of the iris itself, while in blue eyes it arises from the posterior pigment cells showing through the nearly or quite colorless texture of the iris substance. Between the two epithelial layers is a quantity of non-striated muscle tissue, connective-tissue cells and fibres, and pigment cells of various shapes. Arranged as a ring around the pupil is a layer of this muscle tissue about $\frac{1}{45}$ of an inch in width. It is the constrictor, or sphinctor of the pupil. Other muscle fibres commence at the periphery aud converge toward the pupil,—the dilator of the pupil. The iris is very vascular, and well supplied with nerves from the 5th cranial and from the ophthalmic ganglion. They follow the course of the blood-vessels and form a fine plexus at the margins of the pupil.

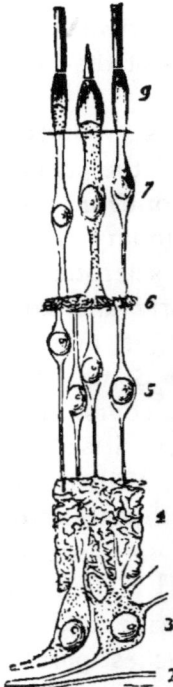

FIG. 162. Nervous elements of the retina.
(Schultze.)

The retina presents an extreme delicacy and a most complicated structure. It extends anteriorly to within $\frac{1}{15}$ of an inch behind the ciliary processes where it terminates in a serrated border—the ora serrata. Near the optic nerve it is about $\frac{1}{50}$ of an inch thick, near the middle $\frac{1}{150}$ of an inch, and near its anterior edge $\frac{1}{250}$ of an inch. At the entrance of the optic nerve is a round disc, porus opticus. It forms a rounded elevation on the inner surface of this membrane, pierced in its centre for the passage of the vessels of the retina. This papilla is known as the colliculus nervi opticus. It is about $\frac{1}{8}$ of an inch within, and $\frac{1}{12}$ of an inch below the antero-posterior axis of the eyeball. In the direct axis of the eye is a somewhat elliptical depression of a yellowish color and called the macula lutea. In the centre of this is a depression, the fovea centralis. The retina may be divided into

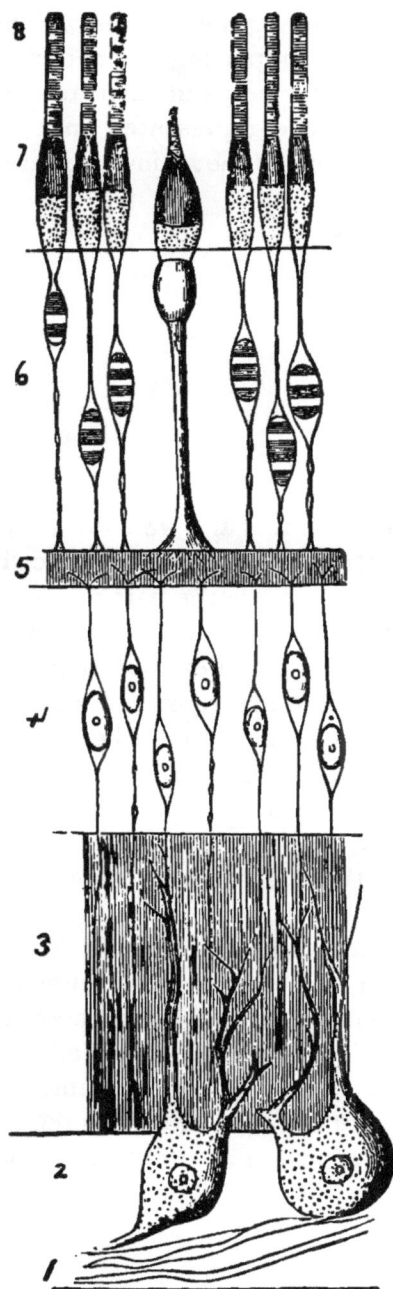

FIG. 163. The nervous and epithelial elements of the retina (semi-diagrammatic.) Quain after Schwalbe. 1, the layer of nerve fibres. 2, the layer of nerve cells. 3, the inner molecular layer. 4, the inner nuclear layer. 5, the outer molecular layer. 6, the outer nuclear layer. 7, the layer of rods and cones. 8, hexagonal pigment cells (not shown.)

eight layers, not counting the two limiting membranes which are more directly connected with the supporting connective-tissue frame-work. In a section made perpendicular to the surface, the retina presents an appearance represented in fig. 162 from M. Schultze. The layers from without inwards are as follows :

1. The layer of pigment cells.
2. The layer of rods and cones.
3. Outer nuclear layer.
4. Outer molecular layer.
5. Inner nuclear layer.
6. Inner molecular layer.
7. Layer of nerve cells.
8. Layer of nerve fibres.

The cells of the first layer are hexagonal and were formerly described as belonging to the choroid. That part of the cell towards the choroid is perfectly smooth while the part towards the rods and cones sends down numerous fine processes. These processes are filled with pigment granules, but they do not completely fill all the spaces between the rods and cones, for, according to Schwalbe, many of the spaces are filled with a liquid substance. The mass of the pigment granules is situated in the inner part of these cells; the part next the choroid being nearly, if not entirely, free from them.

The second layer is known as the stratum bacillosum and it represents the terminal cells of the optic nerve.

It is composed of two elements, rods and cones. With one exception, to be described later, the rods are more numerous than the cones, the latter becoming less and less numerous as the periphery is approached. The rods are long cylinders, in length equal to the thickness of this layer and in diameter about the $\frac{1}{14000}$ of an inch. They are composed of two regular parts, an outer and an inner segment. The outer segment refracts the light more strongly than the inner, which is paler and more granular. The former will not take staining, the latter stains readily with carmine, iodine, etc. The outer segment breaks up into a number of superposed transversed discs only $\frac{1}{80000}$ of an inch thick. On this segment can be seen a longitudinal striation, due to fine longitudinal grooves or depressions. This segment penetrates into the pigmentary layer with a rounded point. Ritter says he has seen a primitive nerve fibrilla in the axis of the rods. At the outer border of the inner segment, and forming a part of it, is a plano-convex body, the "lentiform body," the "rod-ellipsoid" of Krause. The inner segment terminates below in a long, pointed filament, a primitive nerve fibril.

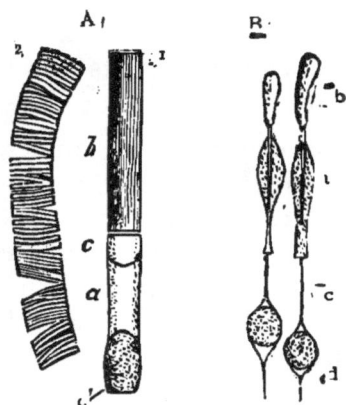

FIG. 164. Rods of the retina. From the monkey. A, rods, after maceration in iodized serum, the outer segment (b) truncated, the inner segment (a) coagulated, granular, and somewhat swollen. c, filament of the rods. d, nucleus. B, rods from the frog. 1, fresh, magnified 500 diameters. a, inner segment. b, outer segment. c, lentiform body. d, nucleus. 2, treated with dilute acetic acid and broken up into plates.
(Schultze.)

The cones are about one-half the length of the rods and are similar to them in structure. The outer segment is known as the "cone-rod," and it possesses in common with the rods a tendency to break up into transverse discs, although this tendency is not so great as in the rods. The inner segment or cone-body is like the rods, longitudinally striated. In the cone-body a structure is found identical with the rod-ellipsoid of Krause.

These appearances in both rods and cones are due probably to this fact: the upper part only of the inner segment is longitudinally striated, for the lower part is homogenous.

In the case of the rods the fibrillated part of the inner segment occupies about the outer one-third, while in the cones it occupies as much as the outer two-thirds. The outer parts are strongly refractile, which fact does not appear to depend in the least on the degree of fibrillation of the segments, for sometimes the outer part of the inner segment of the cones appears entirely free from fibrils, yet it is easily distinguished by its great refracting power.

The third layer consists of granules or nuclei in connection with the delicate prolongations of the rods and cones. The granules of the rods are swellings on the prolongations, one to each fibre. They are situated some distance from the bases of the rods, near the molecular layer. Each enlargement has a nucleus and is characterized, according to Henle, by a cross-striped appearance. The rod fibril prolonged from this granule is interrupted by numerous varicosities and finally terminates in a larger varicosity just before entering the molecular layer. The granules of the cones are not crossed by any bands. Each granule has a nucleus and nucleolus and the whole cone fibre is much larger than that of the rods. The fibre terminates below in an expanded extremity from which proceed numerous prolongations into the molecular layer.

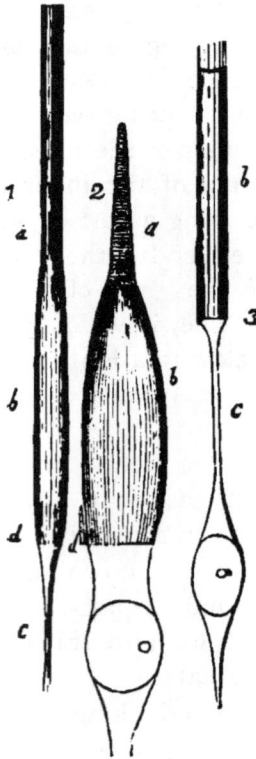

FIG. 165. Fibrillated covering of the rods and cones. 1, rods. 2, cones of man. a, outer, b, inner member. c, rod-filament. d, llmi-tans-externa. 3, rod of the sheep. The fibrillæ project beyond the inner member. The outer member is wanting. (Schultze.)

In the molecular layer little is to be recognized but a granular appearance, minute fibrillæ and a few nuclei.

The inner nuclear layer consists mainly of nucleus bodies, like those of the outer nuclear layer, only larger.

In the layer of nerve cells are found large multipolar cells resembling those found in the brain. They measure from the $\frac{1}{2500}$ to the $\frac{1}{750}$ part of an inch in diameter. Each cell has an unbranched process, axis cylinder, which passes inwards among the fibres of the inner layer, with one of which it doubtless becomes continuous. The branched processes extend into the outer layer and are soon lost in its substance. As a rule they exist in a single layer, but near the yellow spot they are from five to ten layers deep. The eighth layer is composed of the most minute nerve fibrils varying from $\frac{1}{50000}$ to $\frac{1}{25000}$ part of an inch in diameter. The delicate structures of the retina are held in place by a system of connective-tissue fibres, the "supporting fibres of Müller." These fibres commence just beneath the inner retinal layer, where by connected bases they form a boundary line,—the membrana limitans interna. The fibres then pass through the several layers, as illustrated in the figure, until they reach the bases of the rods and cones, where they form a boundary line also,—the membrana limitans externa. From here branches pass between the rods and cones and invest their bases.

MACULA LUTEA.

The macula lutea is an oval spot about $\frac{1}{12}$ of an inch in its horizontal diameter. Its yellow color is due to the presence of a peculiar yellow pigment, which is not deposited in grains. It is a diffuse hyaline coloring matter, soluble in water or alcohol. The macula lutea presents the following histological peculiarities :

1. The layer of nerve fibres is wanting.

2. The layer of ganglionic nerve cells is increased from a single layer to six or eight layers ; when they reach the fovea centralis they are entirely absent.

3. Several of the remaining layers diminish in thickness as they approach the yellow spot and disappear at the fovea centralis.

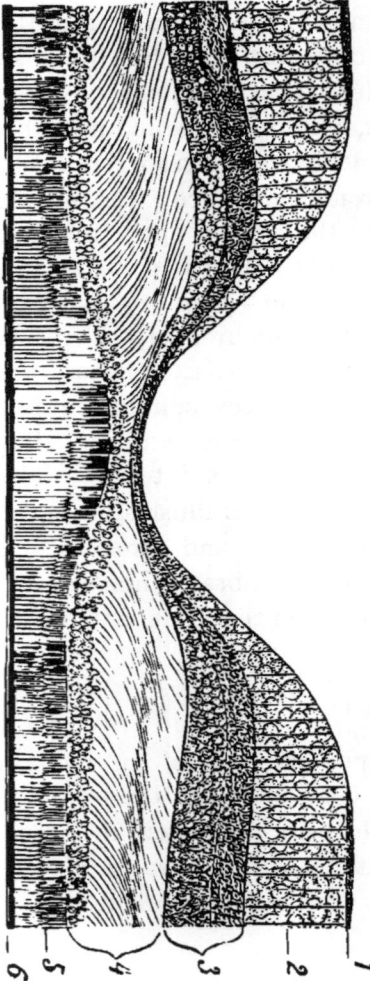

4. As a result of these changes only two layers remain at the central depression, the outer nuclear layer and the layer of rods and cones. Even these exhibit peculiarities.

5. The fibres with which the nuclei of the third layer are connected, are arranged in an oblique direction, reducing the layer much in thickness. At the borders of the macula lutea the cones are separated from each other by a single layer of rods, while in the fovea centralis the rods are altogether absent. Here the cones are longer and more slender than elsewhere.

FIG. 166. Diagrammatic section of human retina through the macula lutea and fovea centralis. 1, internal surface of the retina, in contact with the vitreous body. 2, ganglionic layer of nerve cells. 3, intermediate layers of the retina, disappearing at the centre of the macula lutea. 4, layer of nuclei, showing the oblique course of the fibres in this region. 5, layer of rods and cones, consisting at its central portion exclusively of attenuated and elongated cones. 6, external surface of the retina, in contact with the choroid. In the middle of the diagram is the depression of the fovea centralis. (Schultze.)

VITREOUS HUMOR.

The vitreous humor occupies about four-fifths of the eye-ball and is of a soft gelatinous consistence. A concavity exists in its forepart for the reception of the lens and its capsule. When treated with certain reagents it has the appearance of being composed of distinct membranes, arranged concentrically, and between these membranes there is a fluid substance. A radial striation has been observed in the human vitreous body. Very nearly in the antero-posterior axis of the eye is a canal, lined with a distinct membrane and filled with a fluid ; it extends from the optic nerve to the posterior capsule of the lens. This is the "canal of Stilling." In the vitreous body are corpuscles resembling white blood-corpuscles.

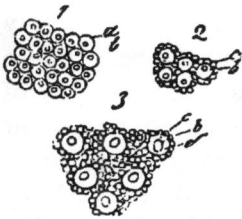

FIG. 167. The rod layer seen from without. a, cones. b, cone rods. c, ordinary rods. 1, from the macula lutea. 2, at the margin of the same. 3, from the centre of the retina. (Helmholtz.)

The hyaloid membrane is an exceedingly delicate membrane, measuring about $\frac{1}{8000}$ of an inch in thickness. It is in contact with the retina externally and with the vitreous body internally. When near the ciliary processes it divides into two parts. The posterior portion lines the concavity in the forepart of the vitreous body. The anterior divides into two parts. 1. The anterior of these parts is known as the zone of Zinn ; it terminates in the anterior capsule of the lens, and it is thrown into folds or plaits that correspond with the folds of the ciliary processes into which they fit closely. 2. The posterior of these parts is very thin and is attached to the posterior capsule of the lens. These two parts by their division enclose a triangular space, called the "canal of Petit." After death this canal is found filled with a serous fluid.

THE LENS.

The lens is a transparent bi-convex body placed directly

behind the pupil.   It is completely surrounded by a thin membrane, the capsule of the lens,—which is lined with a layer of
delicate cells.

The lens is about one-third of an inch in its transverse,
and one-fifth of an inch in its antero-posterior diameters.   Its
posterior surface is more convex than its anterior.   When
viewed with a low power the lens presents a star with from
nine to sixteen radiations.   In the fœtus there are but three
radiations upon either surface.   The rays of the stars of one
side are situated between the rays of the other side.   The
outer portions of a fresh lens are much softer than the inner
portions and they are easily detached from them; the hard
centre is known as the nucleus of the lens.   The lens is composed of a great number of six sided prisms arranged closely
together with but little intervening cement-substance.   These
fibres are about the $\frac{1}{4100}$ of
of an inch broad and their
edges are many times quite
regularly dentated.   They
pass in a curved direction
from the centre and from
the rays of the star to the
periphery where they turn
and pass over to the other
side to its star.

The fibres of the super-
ficial layers have an oval nucleus and sometimes a nu-
cleolus, proving that all the
fibres of the lens must be regarded as elongated cells.

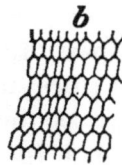

FIG. 168. A, longitudinal view of the fibres of
the lens from the ox, showing the serrated edges.
(Quain.) B, transverse section of the fibres of the
lens from the human eye. (Kölliker.)

In the more internal parts of the lens the nuclei are absent
and the fibres are harder.   The capsule of the lens is a very
thin, transparent membrane, about $\frac{1}{2000}$ of an inch thick at

its centre and so elastic that when ruptured it will frequently contract with sufficient force to expel the lens body. It is lined anteriorly with a layer of nucleated polygonal cells averaging $\frac{1}{1500}$ of an inch in diameter. No such cells line the posterior part.

METHODS OF EXAMINING.

To understand the relations of the different parts to each other, sections of the whole eye should be made by the aid of the freezing microtome.

The cornea may be studied in many different ways. The different layers should be examined first. For this purpose the cornea is removed from the eye of an ox and placed at once in Müller's fluid where it is allowed to remain for two weeks, when it is transferred to alcohol. In two or three days sections can be made vertical to the surface, stained in hæmatoxylin, cleared in oil of cloves and mounted in dammar. To see the branching and anastomosing connective-tissue corpuscles the cornea is placed in a .½ per cent. chloride of gold solution until the tissue is of a pale, yellow color; after washing it is transferred into water slightly acidulated with acetic acid. Upon exposure to light the specimen turns to a violet color when it is examined with a high power. If the epithelial cells render the specimen indistinct they are removed by brushing or scraping. After the

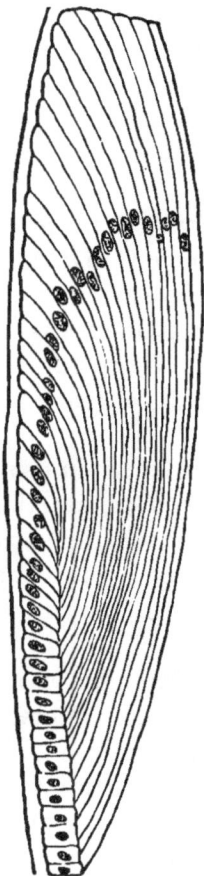

FIG. 169. Section through the margin of the rabbits lens, showing the transition of the epithelium into the lens fibres. (Babuchin.)

gold staining the corneal corpuscles may be isolated.   For this purpose a 25 per cent. solution of sodic hydrate is used ; this destroys the intermediate substance before it does the corpuscles.   The cornea, deprived of its epithelium is placed in this solution for 30 or 40 minutes, when the alkali is replaced by water, to which has been added a few drops of acetic acid. If portions of this membrane be examined in glycerine the corpuscles will be displayed beautifully ; the delicate nerve fibrils and the anastomosing branches of the cells will be clearly brought out.

Nitrate of silver staining can be employed and the tissue treated as usual in such cases.

To examine the retina an unopened eyeball is placed in Müller's fluid for two weeks and then transferred to alcohol for a few days.   Sections of the retina made vertical to the surface can be examined in glycerine or first stained slightly with carmine.   Sections should be examined from different parts of the retina when the local changes will be made apparent.

Sections should be made from a retina hardened in osmic acid.   For this purpose small pieces are removed as carefully as possible and placed in a 2 per cent. solution of this acid. Here they should remain for six or eight hours, when they should be washed thoroughly in water, stained with hæmatoxylin, embedded, sections cut and examined.

To macerate the retina and thus procure its elements in an isolated condition several reagents are employed.   Notably the 2 per cent. osmic acid solution.   After remaining in this solution for six or eight hours a piece of the retina is transferred to a glass slide and by the aid of needles carefully teased in dilute glycerine ; or the acid may be reduced onehalf and the tissue allowed to remain in it 24 hours.   The retina is suitable for teasing just after removal from Müller's fluid.

The retina in the fresh condition should be examined. As soon as removed from the eye a piece of it is teased with needles in a drop of the aqueous humor, covered, and examined with a high power.

The lens is hardened in Müller's fluid and alcohol for obtaining sections. To isolate the fibres it is boiled for ten minutes in a 1 per cent. solution of sulphuric acid and then teased in glycerine. If too transparent slight tingeing with aniline blue will be suitable. Many times very instructive sections are obtained from a lens that has been exposed to the air on a glass slide for a day or two. It has acquired such a degree of consistence that sections can be cut conveniently in the direction to show the delicate hexagonal cut ends of the lens fibres. Other parts of the eye should be hardened in Müller's fluid and alcohol for examination ; sections made and stained as with the simpler tissues.

To study the vascular arrangement, the eye of an ox is carefully removed from its bony socket and injected from the artery with the Prussian blue.

For general purposes of study the human eye is preferred if it can be obtained perfectly healthy and in a fresh condition. If not, the eyes of the pig, albino rabbit and ox will be found very suitable.

CHAPTER XXI.

Tumors.

HYPERTROPHY may be defined to be an increased nutritive activity of a part. It may be either simple or numerical. By simple hypertrophy is understood an enlargement of a part by the increase in size of its anatomical elements. In numerical hypertrophy the enlargement takes place through an increase in the number of cells. These are generally associated.

Hypertrophy is most frequently met with in the muscular system where it is often conservative in nature, as when the walls of the bladder are increased in thickness to give additional power to overcome some obstruction at its base, *e. g.* enlarged prostate, etc., or when the walls of the intestine become thickened above a point of obstruction, or when the gastrocnemei become largely developed in the ballet dancer to overcome the extra strain put upon them. This increased activity may give rise to entirely new elements—to new formations. The inflammatory new formations are very unstable and when their cause, usually some irritation, is removed, they will have a strong tendency to return to a healthy standard or condition. The non-inflammatory have great independence, grow by an inherent activity of their own, and are constantly tending to become removed farther and farther from a healthy condition. Their general tendency is to increase in size, although after a time they may remain permanent.   To this class belong the new formations, known as tumors.   A tumor is many times pathological simply because its specific elements occur in a place where they do not normally belong.   Virchow calls a tumor composed of but one tissue "histioid," when composed of several tissues "organoid."   When in addition to the latter there are organ-like tissues " systematoid."

If a new formation occurs in a tissue agreeing with it in structure it is said to be " homologous."   If unlike it, it is "heterologous."

All of the pathological elements found in a new formation, including the cells, nuclei, matrix, vessels, etc., are prototypes of those found in the normal tissues, only undergoing change and destruction more readily.

The cells of these growths are reproduced most frequently by cell division, the nucleus dividing first, followed by a division of the formed part of the cell.   This has been observed to occur in a very few seconds.   Sometimes the nucleus alone will divide, these nuclei thus formed dividing again and again, until one cell may possess in this way from four to twenty or more nuclei.

These cells are known as the " giant," " mother," or "myeloid" cells.   They are found normally in the medullary substance of bone.

Pathological cells then come from pre-existing cells, and when newly formed are usually small and round, having a nucleus and also nucleolus, or composed of nucleus matter alone, simple undifferentiated protoplasmic cells.   At this stage it would be impossible to tell the future of the growth.   Like the small cells of the embryo, they are entirely undifferentiated. These cells may be the round cells of a sarcoma or the cells of connective tissue.

As soon as a tumor is completely developed it is liable sooner or later to undergo some of the forms of degeneration.   If it has been of short duration, attained a considerable size, and if it is composed largely of cells, then it will undergo these changes all the more rapidly.   If it has been of slow growth and its elements are developed into tissue, then it will not be liable to degenerate.   Fatty degeneration is most commonly met with. This is probably due to the fact that in the rapid formation of new tissue there is not a proportionately new formation of blood-vessels, and as a result of the insufficient circulation and want of nutritive material, the fatty metamorphosis occurs.

Tumors may also undergo pigmentary degeneration, usually from a deposit of melanin.   This is a black, or nearly black, substance found physiologically in the skin and eye.   It is seen either as free granules in the tumor or deposited in the cells.   It does not appear to be at all susceptible to reagents, and its origin is probably the same as that of hæmatoidin.   In caseation the fluids are absorbed and the elements are dried up, changed into a yellowish cheesy material, which process may continue until the whole mass may become surrounded by a capsule of fibrous tissue.   In calcification, small calcareous particles are infiltrated through the mass.   Sometimes softening liquefies the whole mass into a thin liquid, which under the microscope is seen to consist of broken down material, granular matter, fat, etc.

Colloid and mucoid degenerations also occur when the albuminous ingredients are transformed into substances chemically resembling mucin and an allied colloid material.

A tumor is malignant when it has a tendency to recur in the same or some distant place after its removal.

It is innocent when this tendency is not present.   The term "malignancy" then, is purely a clinical one and does not refer to any property of the growth to destroy life.   The heterologous character of a growth is an evidence of its malignancy.

In the examination of tumors the fresh cut surface should

be scraped and this examined for cells, their shape, number, size, nuclei, the size and number of nuclei in each cell, all should be carefully noted. Then the tumor should be cut in small pieces not over one-half an inch square and placed at once in dilute alcohol, to be replaced in a few days by common methylic alcohol, and if the tissue still remains too soft for cutting thin sections, stronger alcohol may be added for a day or two. Müller's fluid may be employed, but at this laboratory the best results have been obtained by the use of alcohol alone. In two weeks the tissue will be of sufficient consistence to allow thin sections to be made with the aid of a razor. By holding the piece of tissue firmly between the thumb and fingers of the left hand, the razor held in the right hand can be drawn from heel to point over the tissue cutting the section sufficiently thin for examination. Or by using one of the embedding mixtures already given the piece may be embedded in the microtome and sections cut as has been described. The arrangement of the fibres and cells should be noticed together with any alveolar stroma that may be present.

Carmine and hæmatoxylin are useful staining agents.

If it is desired to secure the specimen permanently it should be cleared in the oil of cloves and mounted in dammar.

It is very difficult to make a satisfactory classification of the new formations. The classification given in T. Henry Green's "Pathology and Morbid Anatomy" is as free from objections as any with which we are acquainted. It is here given with slight changes :

CLASSIFICATION OF TUMORS.

I. Type of the fully developed connective tissues.
      Type of fibrous tissue, - - - Fibroma.
      Type of adipose tissue, - - - Lipoma.

Type of cartilage tissue, - - - Enchondroma.
Type of bone tissue, - - - - Osteoma.
Type of mucous tissue, - - - Myxoma.
Type of lymphatics, - - - - Lymphoma.
II. Type of higher tissues.
Type of muscle, - - - - - Myoma.
Type of nerve, - - - - - Neuroma.
Type of blood-vessels, - - Angioma.
Type of papillæ, - - - - - Papilloma.
Type of secreting glands, - - Adenoma.
III. Type of embryonic tissue.--The sarcomata.
Spindle-celled sarcoma.
Round-celled sarcoma.
Myeloid sarcoma.
IV. The carcinomata.
Scirrhus.
Encephaloid.
Colloid.
Epithelioma.

FULLY DEVELOPED CONNECTIVE TISSUE.

Of the two kinds of corpuscles found in connective tissue, the movable or wandering kind is the most important in this connection. In size, contractility, ability to wander, etc., they seem identical with the white blood corpuscles and pus corpuscles. In all probability they have their origin in the blood. It is not known whether they can pass into the regular connective-tissue corpuscle or not. Neither is it known in what channel they move.

TYPE OF FIBROUS TISSUE.

Fibroma, fibroid or connective-tissue tumor. This tumor consists of quite distinct fibres that are without any arrangement, and separated only with difficulty. If the section be made across a blood-vessel the fibres will be seen running a circular manner around it, as seen at the left upper corner of figure 170. Only a few cells will be found, and these are most abundant in the neighborhood of blood-vessels. They are usually of the spindle-shaped or stellate variety. Nuclei that take the staining readily are seen distributed over the field. As a rule there are but few blood-vessels, but it sometimes occurs that the walls of the vessels have become firmly united with the structure of the tumor, hence if the growth be cut into or injured severely the mouths of the vessels will not be able to contract and profuse hemorrhage results.

FIG. 170. Fibroma.

In size the fibromata vary from a very small circumference to an immense growth. Their form is also varying. The fresh cut surface is usually dry, only in the rapidly growing younger growths when a serous or mucous fluid exudes. Arising from the skin they are usually softer and less dense than those found in other parts, and in this situation are usually single. They are generally limited by a capsule and have a slow growth, occurring in middle and advanced life. They increase in size by a central growth, by a multiplication of their own elements, and do not invade the surrounding healthy structure.

They are then innocent growths and cause disturbance to the organ or tissue in which they are situated and to the whole organism only from their size. The fibromata are not liable to undergo degeneration. Fatty degeneration, calcification, mucoid softenings and hemorrhages are met with usually affecting only a part of the growth. Growing beneath the skin these tumors are sometimes soft, without a capsule and

multiple.   They are known here as wens.   Nasal polypi are a
variety.   So is a tumor often described as a neuroma, which
under the microscope is seen not to consist of true nerve
tissue, but is fibrous.   These growths usually commence from
the connective-tissue surrounding the nerve, the neurilemma,
and by increasing in size, either press upon the nerve proper
or grow around it and thus as they increase in size they compress the nerve.   They are generally small, round, hard
tumors that are painful in the extreme.   Uterine fibroids are
rarely composed of fibrous tissue.   They will be described
under muscular tumors.   The fibromata are frequently combined with other forms.

TYPE OF ADIPOSE TISSUE.

In structure a lipoma resembles ordinary adipose tissue,
consisting of large cells that are fully distended with fat.   The
nuclei of the cells are not visible unless the fat be dissolved
from the cells, or unless a cell is found containing but very
little fat.   They vary in size, frequently attaining a most enormous growth.   The fresh cut surface shows fatty tissue.   It
occurs most frequently in parts where fat normally exists,
rarely in other parts, is usually sharply circumscribed, encapsuled, grows slowly and with a central growth.   It has no tendency to return after removal.

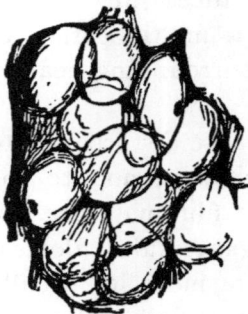

It rarely undergoes any of the degenerations and when occurring only small
parts are affected.   Figure 171 shows a
thin section of this tumor with the inter-
cellular connective-tissue.

TYPE OF CARTILAGE.

FIG. 171. Lipoma.

Enchondroma, chondroma.   This tumor is rarely found composed of cartilaginous tissue alone but usually combined

with connective tissue. It may be either hyaline, reticular or fibrous cartilage, or all three combined. The number and size are very variable. Some are spindle-shaped, some stellate and movable. Usually, however, they resemble the cells of normal cartilage. The enchondromata vary in size, are usually single, occasionally multiple. They occur in the early part of life, even in the new born. By far the greater number affect the bones and most frequently the medulla. Thus the articulating surfaces are rarely affected. They may arise from cartilage itself, their likeness to normal cartilage is then more exact. In some of the softer forms there is a tendency to return after removal, affecting even the lymphatics, and in the young causing cachexia. The malignant properties of the enchondromata, when present, are probably due to the fact that sarcomatous elements are associated with them. However healing almost invariably occurs after complete extirpation, and in the case of a pure enchondroma malignancy may be said to be entirely absent. Of the many degenerations to which this tumor is subject calcification is the most common. Ossification sometimes affects the periphery of the growth so that it is surrounded by a thin bony wall. Spiculæ of bone are frequently found through the growth. A specimen in the author's possession shows about one-third of the growth truely ossified, the remainder resembling normal cartilage. The line between the two being sharp and distinct.

For an illustration of this kind of growth see cartilage.

TYPE OF BONY STRUCTURE.

Osseous tumor, Osteoma. In the case of this tumor the bony appearance is the natural result of development, whereas in many other cases—tumors having undergone osseous degeneration—it is accidental. It has an independent growth and is not to be confounded with the products of inflammation

of bone, as the callus after fractures, etc.   Most of the osteo-
mata arise from connective-tissue.   They may have their
origin from cartilage bone, or the periosteum of bone.   Those
having their origin apart from bone, heterologous, are known as
osteophytes.   They are found near diseased joints, near the
seat of inflammatory processes and in many other situations.
They are found not uncommonly in the lungs and brain.
They are to be carefully distinguished from growths that have
become partly ossified, for in the latter case they might be
more or less malignant, while a true osteoma is perfectly innocent.   The homologous exostoses are found most frequently
on the external and internal surfaces of the skull, in the orbit,
on the upper and lower jaw, etc.   They are troublesome only
when some neighboring part is affected by pressure.   The
appearance under the microscope is not unlike that of the true
bone, at least the lacunæ and canaliculi, are present, although
not arranged in any order.

TYPE OF MUCOUS TISSUE.

Myxoma, mucous tumor, tumor mucosus, gelatiniform or
colloid sarcoma.   A myxoma con-

FIG. 172. Myxoma.

sists of a mucous basis substance in
which are spindle-shaped or stellate
cells which anastomose with each
other.   Some of these cells are
shown in figure 172.   A few are
round or oval or spherical.   This is
very generally the case in the
younger growths.   If young and
rapidly growing the number of these cells will be largely increased proportionately.   A nucleus is seen in each of the cells.
Sometimes two nuclei are present.   The refracting power of
the mucus is so great that some care is necessary in order to
see the outlines of the cells.   Staining will be of advantage

here.   The cells are easily obtained by simply scraping the cut surface and adding a little saline solution to the scrapings. They are closely related to cells found in the sarcomata, and by many are so classed.   The same kind of tissue exists in two places in the body physiologically, in the vitreous humor of the eye and in the umbilical cord.

The myxomata usually occur as single tumors, and are generally round, uniform and small.   The fresh cut surface may show septa of connective tissue, giving the growth a soft but quite firm consistence, or the connective tissue may be nearly, if not entirely absent.   There will then escape a viscid mass of mucilagenous consistence to such a degree that the whole tumor will become flattened and formless.   Their most favorite seat is in the adipose tissues and they are here generally encapsuled.   Their growth is usually slow although they are many times of extraordinary size.   The walls of the bloodvessels are very thin and liable to rupture.   Hence the frequency with which sanguineous cysts are met with.

The cells themselves may become destroyed by either fatty or mucoid degeneration.   As a rule the myxomata are

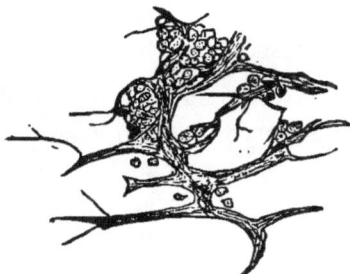

FIG. 173.   Lymphoma.

innocent growths.   Sometimes, however, they exhibit malignant properties.   This probably is due to the fact that many times these growths are combined with others, especially the sarcomata.   Figure 172 represents some of the cells of a myxoma removed from the vagina.   The growth was about the size of a walnut.

TYPE OF LYMPHATIC TISSUE.

Lymphoma, Lymphadenoma.   Figure 173 represents a section of a lymphoma of the arm.   It is not very unlike a lymphatic gland in structure, consisting of a basis of distinct fibres which branch and cross each other like a net-work, and of cells identical with the white blood corpuscles. These cells fill up the space in the basis, but in the figure they have been nearly all removed by brushing the section with a camel's hair brush moistened in water. The firmness of the tumor will depend upon the comparative amount of basis fibres and nucleated cells. If the growth is young and increasing rapidly in size then the cells will be the more prominent part of the growth. Later the number will diminish and the reticulum become thicker and firmer. These tumors not infrequently acquire a large growth even infiltrating the surrounding tissues. They are homologous primarily and become heterologous only from the new tissue extending into surrounding parts, or from their growing in a place where the lymphatics are very small and few in number.

The lymphomata are innocent growths and are not liable to undergo degenerative changes. In the disease known as "Hodgkin's disease" the new growths in various parts of the body are like the one described. The enlargement of the spleen in leukæmia is of the same nature.

TYPE OF MUSCULAR TISSUE.

Myoma. A tumor composed of striated muscle is one of the rarest of the new formations. A myoma composed of smooth or non-striated muscle is most frequently met with in the uterus, where it is generally known as a "uterine fibroid," and when projecting into the cavity of the uterus or extending by a pedicle out of the neck is called a "uterine polypus." The muscle cells form one but not the only element. Connective tissue may exist in great abundance. This is especially the case in the older growths. In the new growths it is not uncommon to find almost exclusively the characteristic non-striated muscle cells. There are few blood-vessels distributed through the connective tissue. These are homologous growths, of slow growth, usually single, but often multiple. They are liable to undergo softening, or more freqently to become calcified. They are perfectly innocent, exhibiting no tendency to return after removal.

TYPE OF NERVOUS TISSUE.

Neuroma. These consist of true nerve fibres and are not the growths so commonly met with growing from the sheath of nerves or within the sheath. They are composed of ordinary medullated nerve fibres associated with connective tissue. They are found on the ends of divided nerves, growing after amputations. They are usually very small nodules, innocent, and are remarkable largely for the great pain they cause.

TYPE OF BLOOD-VESSELS.

Angioma. These tumors are composed of blood-vessels held together by connective tissue. The diagnosis is readily made without the aid of the microscope.

TYPE OF PAPILLÆ.

Papilloma, papillary or villous tumor. This tumor consists of a body of connective tissue with a covering of epithelial cells, resembling the papillæ of the skin. They are rarely without blood-vessels which end either in a capillary net-work work or in single loop. Cells may be seen scattered through the connective-tissue basis. The epithelial covering is generally like that from which the part arises. The papillomata may occur on any surface of the body, but more generally where papillæ and villi normally exist. They occur singly or many papillæ may be affected giving the growth a cauliflower ·appearance. The papillomata exhibit no tendency to return after removal, yet in many ways they may become serious troubles. They are liable to undergo ulceration, followed by

hemorrhage, especially when situated in the bladder and intestine. Warts of the skin, common warts and horny growths
are varieties of this class, so also are the condylomata and
venerial warts. Figure 174 rep-
resents a growth of this character.
This will not be mistaken for
epithelioma, for in the case
of a papilloma, the epithelial
cells are in their normal relations
to the part, they are homologous,
while in an epithelioma the cells
are heterologous.

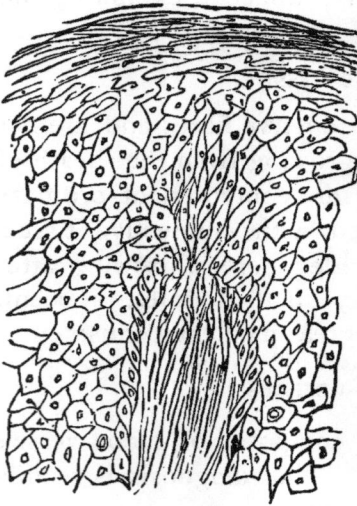

TYPE OF GLAND TISSUE.

Adenoma, glandular tumor. In
structure an adenoma is like that
tissue in which it is found, or
from which it originated, for it
may after a time become completely separated from the old

FIG. 174. Papilloma.

gland. Its function will not be like the normal gland however. Indeed it can be said to have no function whatever.
The adenomata are very difficult tumors to diagnose, being
very liable to undergo the degenerations, especially the formation of cysts and the changing into calcareous forms. They
are frequently, perhaps most frequently, found in the female
mammæ. They are very commonly associated with other
forms as adeno-sarcoma, adeno-myxoma, etc. In the
mammary gland an adenoma is most frequently associated
with a fibroma, giving rise to the familiar adeno-fibroma.
Here the aceni of the gland are separated from each other by
a large growth of fibrous tissue between them, or a bundle of
aceni may be separated from another bundle by an hypertrophy of the intervening connective tissue. This may de-

velop to such an extent that the secreting tubes of the gland will be nearly obliterated. Figure 175 represents the aceni of the gland widely separated from each other.

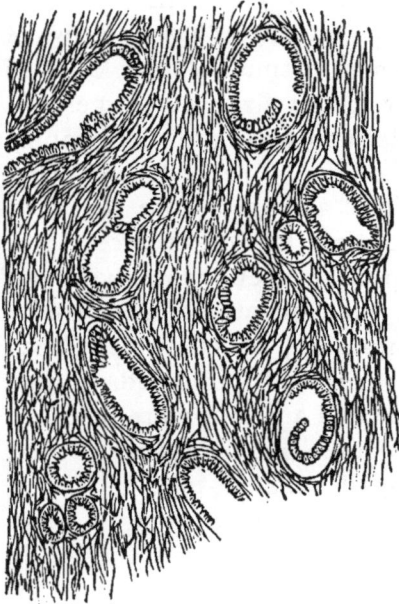

FIG. 175.   Adeno-fibroma, from mammary gland.

The growth of these tumors is usually slow. While they are primarily innocent they may assume malignant properties.

TYPE OF EMBRYONIC TISSUES, THE SARCOMATA.

Fibro-cellular, fibro-plastic, fibro-nucleated, recurrent fibroid, myeloid. The sarcomata are divided into varieties according to the majority of their cells. The spindle-celled sarcoma is composed almost entirely of long fusiform, comparatively thick-bodied, nucleated cells. The processes from either end of the cell are usually long and not infrequently branched. Each cell is possessed with one nucleus frequently with two nuclei. This variety is the most common of this large class of new formations. Figure 176 represents some of these cells taken from the leg of a man. The leg had been diseased nearly two years. A majority of the cells were large, the nucleus multiple in many cells, and a number of free nuclei or small round cells were in the field also. The disease had affected the tibia to its very centre, so that

FIG. 176.   Spindle-cells, from sarcoma of leg.

deep in the growth were found the myeloid cells, seen at figure 177. The spindle-shaped cells vary much in size, both in the same growth and also in different growths. Some growths will be composed almost entirely of cells averaging

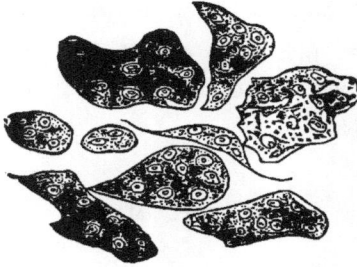

$\frac{1}{1500}$ of an inch length, while others of much larger cells, of twice the size, with larger nuclei, and some growths combine the two. The cells are many times arranged close together, so that there is scarcely any space between them, giving but a small quantity of intercellular substance, which in turn may

FIG 177. Myeloid cells, from sarcoma of tibia.

be either fluid, or granular, or firm and fibrillated. Large cells, large nuclei, and the presence in a cell of more than one nucleus, are evidences of a high degree of malignancy. The cells are not infrequently arranged parallel to each other, running in bundles all through the growth, giving it very much the appearance of a fibroma. This tumor arises as do all the sarcomata, from pre-existing connective tissue, and increases, either by multiplication of its own elements (central growth), or by continually invading the healthy tissue around it (peripheral growth), which is highly characteristic of all this class. The sarcomata are usually quite vascular, the walls of the blood-vessels being composed of embryonic tissue, render them exceedingly liable to rupture, causing the formation of sanguineous cysts, severe hemorrhage, etc. They are also very liable to undergo fatty degeneration. Although this variety may become encapsuled, it possesses unmistakable malignant properties. The growth is usually rapid.

Figure 178 represents the cells from a melanotic sarcoma of the eye. This tumor extended into the vitreous humor from the choroid, was globular in shape and a trifle over one-half an inch in diameter. These cells are mostly spindle-shaped and nucleated, but they now contain a large amount of dark colored pigment melanin, rendering the nuclei obscure, and many times invisible. The large majority of these growths is found primarily in the eye, where this pigment normally exists. They may arise from the superficial integument. Some-times this pigment will be de-posited only in a slight degree, giving the growth a brownish ap-pearance. Then too, only a few of the cells may be thus affected.

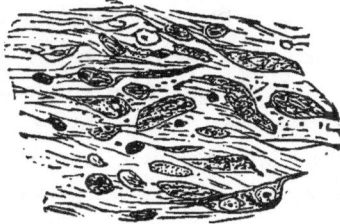

FIG. 178.   Melanotic cells, from sar-coma of choroid.

Again the pigment may be in such excess that the tumor will be a black color. These tumors are very liable to have their elements conveyed to distant parts by the blood-veesels in which case their melanotic character will accompany them. In this way secondary growths are found in the liver, kidneys, lungs, etc. The laboratory is in possession of a liver three-fourths of which has become transformed into little melanotic growths, varying in size from a pea to masses two inches in diameter. This variety of the sarcomata is perhaps the most malignant of all, exceeding in this particular many of the cancers.

An osteoid sarcoma is usually a spindle-celled sarcoma that has either become truly ossified, or more or less hardened by calcareous deposits. It is important to recognize the sar-comatous element, inasmuch as the innocence or malignancy of the growth will depend upon it. Some acid, as dilute hy-drochloric, may be used to dissolve out the calcareous matters when it can be examined for the characteristic cells, which, if found, will decide its malignancy.

Figure 179 shows the large and small round cells of a sar-coma, growing in the orbit after enucleation. This tumor re-

curred after removal, causing the death of the patient.   A sarcoma composed of round cells is usually of much softer consistence than one composed of spindle cells.   Such a sarcoma

FIG. 179. Large and small round
cells from sarcoma of orbit.

is composed of true embryonic connective tissue, with a fine granular inter-
cellular substance.   The smallest cells
take carmine staining evenly, evidently
consisting of nothing but free nucleus
matter.   The   larger   cells   have   a
nucleus while the largest have frequent-
ly two nuclei, with nucleoli.   The cut surface yields a juice
rich in cells.   This variety increases with a rapid growth by
invading the healthy surrounding structures, involving the
lymphatics and internal organs.   It is full of blood-vessels
easily ruptured.   It is not to be mistaken for encephaloid
cancer, which   it resembles by physical characters.   Here
the cells are of a nearly uniform size and character, and
there is an entire absence of an alveolar stroma.   When an
alveolar stroma is present careful attention must be given to
notice whether the cells are grouped together in these alveolar
spaces or exist singly and alone.   If the latter then it is
termed an alveolar sarcoma, if the former, it belongs to
one of the cancers.   It is often very difficult to distinguish between the two.   Figure 177 illustrates the large many nucleated cells of the myeloid sarcoma.   This variety is usually
found growing in connection with bone, especially from the
medullary cavity.   The nuclei vary in number from two or five
to ten or fifty.   These large cells are generally separated from
each other by a number of cells of the spindle-shaped variety,
among which are seen a few round or oval ones.   It is quite
frequently encapsuled, most frequent in early life, and is the
least malignant of all the sarcomata.

Thus it will be seen that all the sarcomata possess malignant properties, in this respect ranking next the cancers.

They disseminate by means of the blood-vessels, and thus rarely infect the lymphatics, a clinical distinction between these growths and the cancers, marked and distinct. For this reason they are reproduced with greater rapidity than the cancers. The lung being the most favorite seat for the secondary growths. No one variety of the sarcomata is necessarily malignant, while again the same variety may recur in the same place many times.

THE CARCINOMATA.

A cancer is a growth consisting of a fibrous, alveolar stroma, the meshes of which are filled with cells of an epithelial type. While the cells have no "specific" character, yet they are recognized by their large size, irregular shape, the prominence, number and size of their nuclei and nucleoli. These cells exhibit every possible shape, as seen in figure 181. They are full of granular matter and from their great liability to undergo fatty degeneration they usually contain some fat globules. In the juice of the cancers will be found numerous free nuclei, especially in the younger and softer growths. Cells not very unlike these are found in the normal tissues or in those tissues when slightly inflamed. The day of the " specific

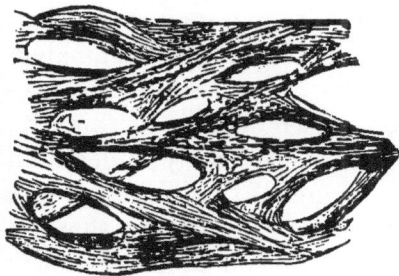

FIG. 180. Stroma of Scirrhus.

cancer cell " is nearly over, in fact there is no such thing at the present time, the very best pathologists hold strictly that "every pathological growth has its physiological prototype."

If cells are found in a growth of the character described above and illustrated at figure 181, then that growth must be looked upon with suspicion, but before pronouncing it a cancer two other things must be carefully noted; first, the stroma, and second, the arrangement of the cells within the alveoli. The stroma, or solid portion of the cancer generally consists of a frame-work of connective tissue as seen at figure 180, so arranged that round or oval alveoli are formed freely communicating with one another, in which are grouped together the cells described above. The amount of stroma varies exceedingly. Sometimes it is so great as to form the largest part of the tumor. The alveolar spaces are then very small, and the growth will be hard to the touch, and the cut surface will yield but little juice. Again it may be very scanty as in the rapid growing and young cancers. Every possible degree as to quantity exists. Cancers have been found in all tissues save cartilage. The female mammæ, uterus, lower lip, stomach, liver, œsophagus and lymphatic glands are all favorite places for the development of the cancers. They may occur alone or in great number, and appear as tumors or as infiltrations. They are very rarely separated from the healthy tissues surrounding them by a capsule, but on the contrary show a close connection with them.

The blood-vessels are arranged very different from those found in the sarcomata. In the latter it will be remembered the vessels ramify all through the growth, and their walls being composed of embryonic connective tissue, they easily rupture and then the elements are easily and rapidly disseminated.

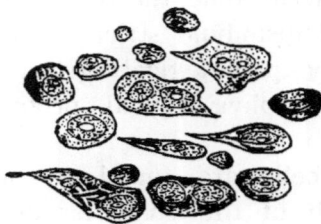

FIG. 181.   Cells from Scirrhus.

While in the former—the cancers— the blood-vessels course within the stroma and very rarely indeed do they communicate with the alveoli. Thus, it is very rare, if ever, that the cancers are disseminated by means of the blood-currents. However a study of the lymphatics shows them to be numerous, accompanying the blood-vessels and communicating freely with the alveoli spaces. The elements thus enter the lymphatics readily and are carried to the nearest glands where they are caught in its meshes and are carried further on, not however until the gland itself has become sufficiently affected to furnish other elements. In the cancers, then, dissemination is slow and accomplished through the lymphatics. All cancers are very liable to undergo fatty metamorphosis, especially the young·and rapidly growing varieties.

Scirrhus or chronic cancer, as its name implies, is of slow growth and of a firm and hard structure. In this variety the alveoli are small and comparatively poor in cells. Instead of the organs affected being increased in size they are many times actually reduced, often depressed in the centre and firmly at-

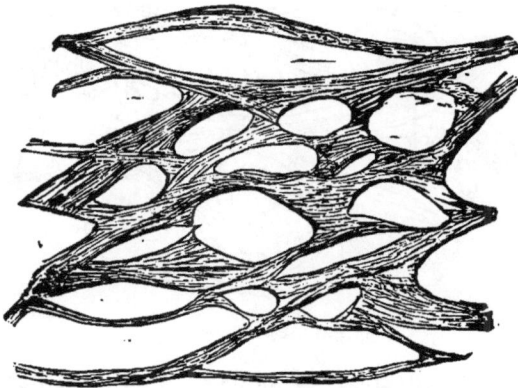

FIG. 182. Stroma of encephaloid.

tached to the skin. A microscopical examination of the centre of this growth may reveal nothing but cicatricial tissue. The cells have suffered degeneration, while the stroma has atrophied and contracted. At the periphery will be found a zone of cells, and free nuclei infiltrating the neighboring tissues. Between the two will be seen the characteristic alveolar stroma as at figure 180, together with the cells as at figure 181. Its most frequent seat is in the female mammæ and in the stomach.   The secondary growths arising from it are generally encephaloid.

Encephaloid or acute cancer differs from the above mostly in its rapid growth and small amount of stroma.   It is usually very soft and by scraping the fresh cut surface an abundance of juice is given off, rich in cells, free nuclei, granular matter, etc.   Figure 182 illustrates the stroma smaller, and the alveolar spaces correspondingly larger than found in scirrhus.   It is not of so frequent occurrence as scirrhus, arising as a secondary growth of the latter.

By colloid is understood a degeneration of the above varieties.   A section of colloid shows a small amount of stroma

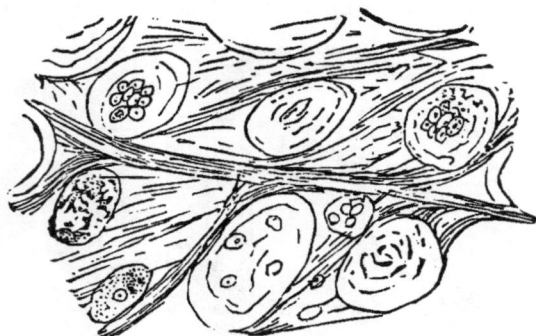

FIG. 182.   Colloid.

and nearly entire absence of cells.   The alveolar spaces are quite well marked, being generally round, varying in size, and filled with the soft, colorless, glistening colloid material, in which are a few cells.   Many times the cells themselves appear filled with this same material.

Epithelioma or cancroid varies much according to its situation.   Arising from the cutaneous or mucous surface the cells will be found to correspond with the cells taken from those surfaces.   Like those found on these surfaces they are usually irregular in shape, containing generally one nucleus, sometimes two nuclei. Their arrangement is most peculiar and characteristic. They appear to arrange themselves in groups and thus they form the "concentric globes" or "epithelial nests." These nests are frequently so large as to be visible with the naked eye, especially when they are of a yellowish color from becoming hard and dry. The epithelium is here heterologous in its nature, extending from the surface into the subjacent connective-tissue, giving the great characteristic of this variety. The point of junction of the cutaneous and mucous surfaces is its favorite seat. Here on the lower lip it is usually seen to commence as a small ulcer, caused by some external irritation, which grows quite rapidly, becoming firm and indurated, with an ulcerating surface. Under pressure the cut surface may yield little worm-like curdy masses, such as can be forced from the sebaceous glands of the skin. By many authors this is considered very characteristic. While all the cancers are highly malignant, some possess this property to a much

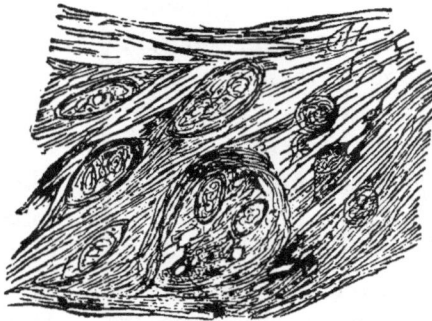

FIG. 184. Epithelioma.

greater degree than others. The vascular and rapidly growing encephaloid reproduces itself in the neighboring lymphatics the most rapidly, while the chronic scirrhus is nearly its equal, colloid is the least so of the three. Epithelioma is by far the least malignant of all the cancers, in this respect ranking below some of the sarcomata. Its thorough removal is not likely to be followed by a return of the growth. It may extend, however, and infect muscle, bone, and lymphatics. It rarely reproduces itself in internal organs, but when it does so the secondary growths correspond to the primary one.

CHAPTER XXII.

Starch.

STARCH is the most generally diffused, excepting protoplasm, of all vegetable substances within the cell-wall. When found in the older structures, roots, stems, seeds, etc., it is found nearly pure; when found in freshly-growing tissue it is in union with chlorophyll. Starch grains contain carbon, oxygen, hydrogen, and some mineral matter. They are insoluble in water, alcohol, ether, and oil; are destroyed by potassa, and colored blue or violet by iodine—the color depending on the density of the granule and the strength of the iodine. The starch grains of different families and different species of the same family differ so much in size and general appearance as to be easily identified. The largest starch grains known are those of tous-les-mois, which are frequently $\frac{1}{300}$ of an inch in length, while the smallest are those of rice, which are occasionally $\frac{1}{1000}$ of an inch in diameter.

Potato Starch.—Botanists have taken the potato-starch grain as the typical form with which they compare others. So we should have a good knowledge of this grain. If the commercial starch is not accessible, the grains can easily be obtained by cutting a fresh potato with a clean knife, and then floating on a glass slide, with a drop of water, the white substance which adheres to the side of the knife. Or, shave off a very thin slice of the potato, and place it in a watch-crystal in a little water; the fine sediment settling to the bottom will be the starch. There are two leading theories regarding their growth. Some claim that the surface of the grain is formed first, and that it grows by layers being deposited on the inner surface of the case, which gradually expands until it reaches its normal size. The other and the more generally accepted opinion is, that the nucleus is formed first, and the grain grows by means of deposits of starchy matter around this nucleus, and each successive layer contains less moisture than the preceding layer; this explains the appearance of rings or laminæ seen so plainly

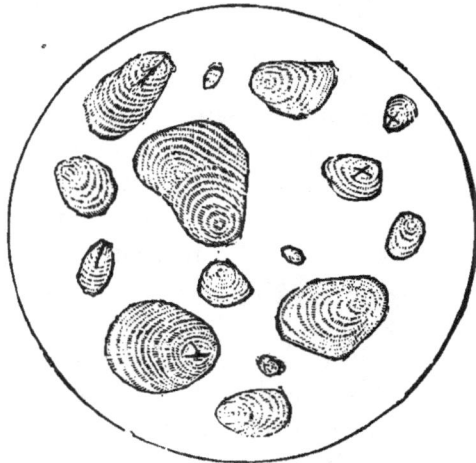

FIG. 185. Potato Starch. x 375.

in the potato and many other starches. A new theory has been advanced in Sachs' Botany (page 59), which is too long, however, for an explanation in this connection. In specimens which have been subjected to even a slight degree of dry heat, there appears a black line or star-shaped mark over the nucleus. The heat evaporates the moisture from the grain, and there must be a shrinkage on the surface to correspond with the evaporation. This is the greatest over the nucleus where is the greatest moisture. The grains are round, ovate, irregularly oval, or egg-shaped, nearly transparent; nucleus eccentric (not in the centre), and in the smaller end of the grain, and surrounded by numerous distinct rings or laminæ. The grains are very irregular in size; the smallest are just perceptible, and the largest are frequently $\frac{1}{400}$ of an inch in length. A very decided cross is seen when viewed with polarized light, the arms of the cross radiating from the nucleus, not from the centre of the grain. This is the cheapest and the most common starch; there being from $800,000 to $1,200,000 worth thrown on the market annually. Probably the greatest part is used for adulterations.

Arrow-root Starch closely resembles potato-starch. The grains are much more uniform in size than those of the potato, and are about $\frac{1}{800}$ of an inch in length. The nucleus is generally in the larger end of the grain, while in the potato-starch, as before mentioned, it is in the smaller end; while the rings are finer and more numerous. Thirty or forty rings can frequently be counted in one grain, while potato-starch sometimes has only three or four. Arrow-root starch takes a distinct cross with polarized light. It is very frequently adulterated with potato starch.

Wheat Starch.—Pure wheat starch can be obtained by cutting through a kernel of wheat, and scraping with the point of a knife a little from the central part of the kernel on a glass slide. There are two distinct kinds of grains found here; small spherical or angular grains floating frequently in a mass, many times more numerous than the large grains, and about $\frac{1}{5000}$ of an inch in diameter. The others are large, lenticular grains, which, when viewed on the face, appear like a spherical grain. When viewed on the edge, they have the appearance of a double-convex lens. This lens shape can easily be proved by touching the cover glass gently with a pencil-point, and watching the grains roll over in the field. This should always be done when testing for adulterations with starch grains. There is seldom any nucleus, but when it is present it is central, and still more seldom are there any rings.   When viewed with polarized light, only a faint cross is seen if any.*   When subjected to dry heat, the grains are changed very much in appearance; being warped considerably from their normal shape. They are larger, more brittle, and more transparent.   Yet generally they can be identified when subjected to either dry or moist heat, if the moist heat be not raised to the boiling point. The large grains of wheat starch in their normal state are very uniform in size for the same variety, but the starch-grains of the different varieties differ considerably in size.   The average

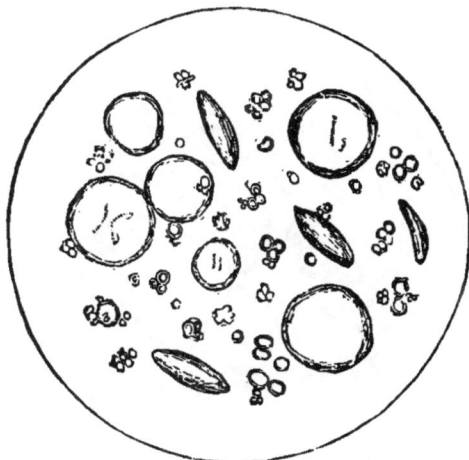

FIG. 186. Wheat starch. x375.

diameter of the grain in the eight varieties examined is $\frac{1}{987}$ of an inch.†   Barley and rye are closely related to wheat.   All of

---

*The statement is made by some botanists that wheat starch gives a cross under polarized light, but I have never been able to detect any closer approach to one than a nearly uniform shadow floating over the grain.

†These measurements have been made with considerable care.   In each case 20 grains, as nearly typical as possible, were selected, and accurately measured; the average was then taken, with the following results: The largest grains of Treadwell wheat measured 1-861 of an inch in diameter; Deihl, 1-816; Wicks, 1-881; Egyptian, 1-994; Russian, 1-1174; Clawson, 1-1256; Schaffer, 1-1000; Vienna flour, 1-861. There is also considerable difference in the size of the small grains. Schaffer small grains measure 1-4700 of an inch in diameter; Treadwell, 1-6102; Vienna flour, 1-5166; Russian, 1-4000; Egyptian, 1-6000.

these are used extensively for adulterations.

Barley Starch is composed of large and small grains. The large grains are smaller than those of wheat; being about $\frac{1}{1600}$ of an inch in diameter. There is less difference between the long and the short diameters than in wheat starch, so that when the grains are rolled over they present less of a lens-shape, being rounder. Rings and a star-shaped nucleus are quite frequently apparent. The small grains are more angular, frequently having a nucleus, and average $\frac{1}{8800}$ of an inch in diameter. No cross is seen when viewed with polarized light.

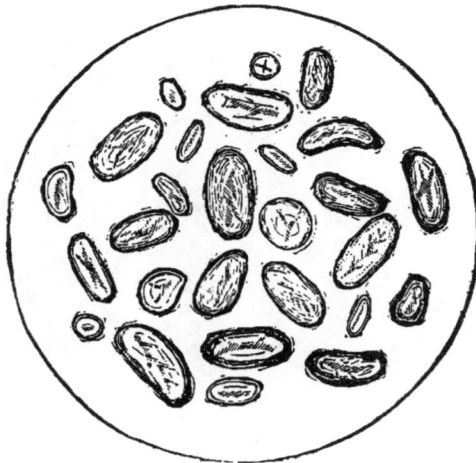

FIG. 187. Bean Starch. x375.

Rye Starch grains are larger than those of wheat, very seldom do they show any rings, and when present they are eccentric; occasionally a star-shaped and central nucleus is present. The large grains average $\frac{1}{727}$ of an inch, the small grains $\frac{1}{5717}$ of an inch in diameter. A distinct cross is seen in rye starch with the polarized light. After examining these starches in their natural condition, they should be subjected to both dry and moist heat, and examined, as their appearance is much changed by heating. As adulterants, they are frequent-ly so treated.

Bean Starch.—We have here a very different appearance from any other starch, excepting that of the pea. The grains are regularly oval and quite uniform in size. A dark line with ragged edges generally extends the whole length of the grain; cross-marks being frequently seen. Faint rings are seen near the edge of the grain. The grains average about $\frac{1}{647}$ of an inch in length and $\frac{1}{1000}$ of an inch in breadth. Dry heat renders the grains more brittle, and destroys the nucleus,

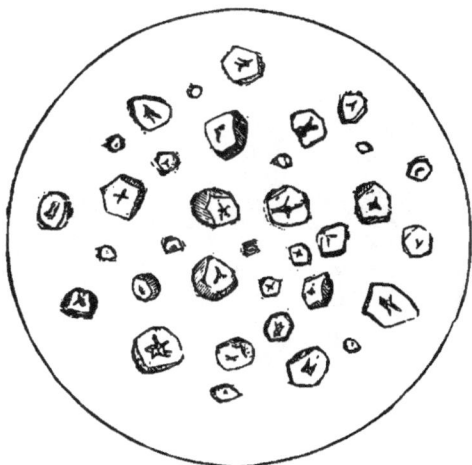

FIG. 188. Corn starch. x375.

but not the rings. Moist heat expands, distorts, renders more transparent, and destroys both rings and nucleus.

Pea Starch is the nearest like that of bean starch. The grains are smaller and more slender, being generally less than $\frac{1}{700}$ of an inch in length.

Corn Starch.—We come now to a starch grain bounded by plane faces and angles instead of curves. The grains are angular, have no rings, and present a round or star-shaped cen-tral nucleus. The average grain is $\frac{1}{1000}$ of an inch in diameter. The shape is only slightly changed by dry heat, but is entirely destroyed by moist heat. The grains found in the central or outer part of the kernel of corn are more angular than those found in the inner part. This variety is frequently substituted for wheat flour, under the name of "amylum."

Rice Starch.—The starch grains of rice resemble very closely those of corn. They are much smaller, however; being only $\frac{1}{835}$ of an inch in diameter. The grains are angular; being bounded by plane sides only, are without rings, and have a central nucleus which is either a dot, a line, or star-shaped. The grains are aggregated together in angular or very ir-

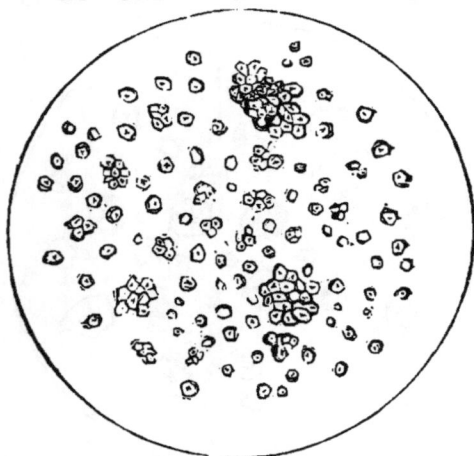

FIG. 189. Rice Starch. x 375.

regular-shaped masses. Rice is used much more extensively in England as an adulterant than in America, and commercial rice flour is frequently adulterated with corn starch.

Oat Starch is the nearest like that of rice, and it is quite difficult to distinguish between them. Oat starch is both compound and simple. The compound grains or masses are oval, spherical, or egg-shaped; the surface of the masses being smooth, while those of rice are irregular. The divisions into grainlets show very distinctly. The simple grainlets are larger than those of rice; being $\frac{1}{4080}$ of an inch in diameter, and

bounded by one or two curved faces. They are without
nuclei and without rings. A faint cross is seen with
polarized light.

Buckwheat Starch is made up of both compound
and simple grains. The compound grains or masses are
cylindrical or prismatic. When cylindrical, the curving surface
is perfectly smooth, but the ends are irregular, as though they
had been broken. These masses are very numerous and characteristic, and somewhat resemble the cell-contents of black

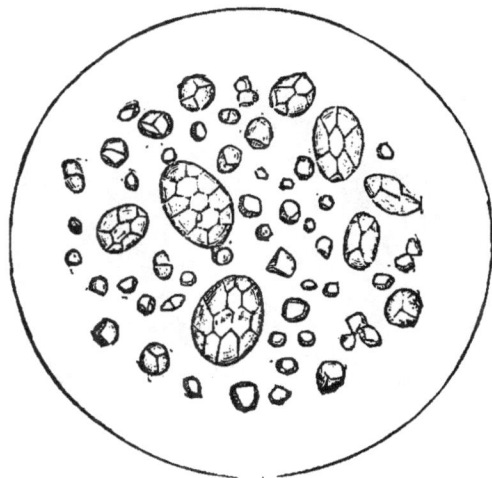

FIG. 190. Oat Starch. x 375.

pepper; being coarser, however than the latter. Black pepper
is largely adulterated with buckwheat. For this reason buckwheat should be compared with some of the grains from the
central part of black pepper, which can be easily obtained by
scraping it out with the point of a pen-knife. The grainlets of
buckwheat starch are like those of rice, in having a central
nucleus and no rings, and are like those of oat, in having one
or more curved faces. In size, they are about $\frac{1}{3000}$ of an inch
in diameter. A correct knowledge of these starches, so closely

related to rice, can be obtained only by faithfully comparing each under the microscope with starch from the latter source.

Sago Starch is obtained from the parenchyma or pith of several different varieties of palms.   Sago appears in market in a variety of forms; as pearl sago, white sago, sago flour, sago meal, etc. When examined with the microscope, the starch grains of sago appear quite large compared with those of the other starches.   They are oval, ovate, or elliptical in shape; much broken, generally one extremity is rounded, and the other extremity, or the sides near it, appear to be clipped,

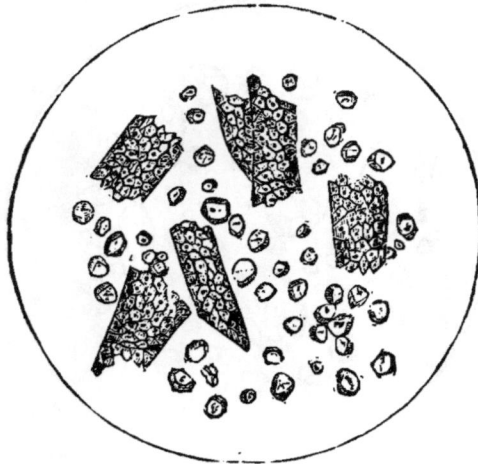

FIG. 191.   Buckwheat Starch,   x 375.

which is due to the pressure of the adjoining starch grains. The nucleus is eccentric, as indicated by a dark cross or slit which frequently extends the length of the grain ; the surface is irregular or tuberculated, and marked by a few distinct rings, fewer than are seen in the potato-starch grain. The grains exhibit a faint cross when viewed with polarized light. The starch grains composing commercial sago are so changed by the process to which they are subjected before being ready for market, that there is little resemblance between them and the fresh grains. The starch grains found in the pearl sago are the most changed by heating. Sago is not used so much in this country for an adulterant as in Europe. Commercial sago is frequently adulterated with potato starch, sometimes with rice. Sometimes there is an entire substitution of potato starch for the sago. Any adulteration used for sago can readily be detected by the microscope, by noticing the above described characteristics.

Tapioca Starch is prepared from manioc or cassava, or, according to Linnæus, from the root of Janipha Manihot. In the preparation of tapioca for market, the substance is subjected to a temperature of 100 degrees C., which changes the appearance of the starch grains very much from what they are in their fresh state, yet they are not entirely destroyed. The heat partially dissolves the outer case of the starch grains, which renders tapioca slightly soluble in water. The grains are quite uniform in size (about $\frac{1}{2000}$ of an inch in diameter); they are round or cup-shaped, with flattenings here and there, due to the pressure of neighboring grains. The starch grains of tapioca are generally found floating in the field singly, but in the growing root they are found compounded of two, three, or four grains each. A distinct and large circular nucleus is seen in fresh specimens. In dried specimens the nucleus is marked by a distinct star or cross. Tapioca is adulterated with rice, sago, and potato starch. Potato flour is frequently prepared like pearl tapioca, and sold as such. Tapioca is used quite extensively in England as an adulterant, but not so much in America. These starches, sago and tapioca, are so much changed in the different commercial varieties, i. e., pearl, white, meal, etc., that to become well acquainted with them one should examine each variety carefully. An illustration or drawing of these in their fresh state would hardly be of value in identifying the starch grains as we find them in market as an adulterant.

Turmeric Starch is from the rhizome of Curcuma longa,

and is imported principally from Southern Asia. The parenchyma is packed full of starch in angular or roundish
masses. Turmeric is used extensively as a coloring material,
to give deeper color to the spices which have been adulterated
with some of the flours. When a ground spice, as, for
example, mustard, contains turmeric, even if not in large quantities, its presence can be detected by exposing the mustard to
the light, when it will fade to a dingy yellow. Its presence
can also be detected by treating the suspected substance with
potassa, and if turmeric be present the substance will turn a

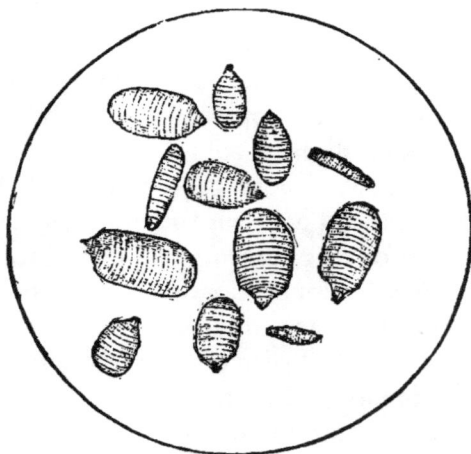

FIG. 192. Turmeric Starch. x 375.

deep yellow or brick-red color. The starch grains are quite
uniform in size, and in shape are elliptical, oval, or like flattened discs, sometimes even truncated. The nucleus is at
one extremity, and has the appearance of being entirely outside of the grain proper. Rings quite distinct, numerous and
uniform in density, pass around the grains like zones, and present a beautiful appearance in a fresh grain. Commercial
turmeric has been heated so much in preparation for market
that frequently the rings cannot be seen, and even the normal

shape of the grain is lost.   In the fresh state they show a decided cross or black bands with the polarized light; but this is seldom seen in commercial turmeric.   The coloring material is a deep, reddish yellow, and is contained in special cells of the parenchyma.   The starch grains are white.   The action of iodine and potassa is the same here as with all starches, but sulphuric and sulphochromic acids are of perhaps more value in this case, for they turn the coloring matter to a peculiar rose-pink.   In the examination of mustard, this test is valuable.   Of the twenty specimens of mustard examined, during the past two years, every one contained turmeric.   It is used to color many other spices.   The turmeric of commerce is itself adulterated frequently with corn starch, etc.

Ginger Starch grains are irregularly spherical, oval, or disc-shaped, closely resembling those of turmeric, belonging to the same family, Zingiberaceæ.   The nucleus is at the extremity, as if it were hardly a part of the grain, the rings are numerous and uniform.   A cross is seen with polarized light.

Much of the ginger of the market has been scalded, which causes the starch grains to lose their normal shape.   It is difficult then to see the rings, and the cross, which was seen with the polarized light, is destroyed.   In examining the starch from the root, as found in the stores, the starch grains at the centre will be found to be more perfect than those taken from near the surface of the root.*

*The following references may be of value to those wishing to carry the study of the starches farther: Hassall's "Adulterations in Food and Medicine;" Sachs' "Botany," page 56 ; Souberian, "Dictionnaire des Falsifications ;" Wiesner, "Rohstoffe des Pflanzenreiches," pp. 239-289 ; Planchon, "Détermination des Drogues Simples," Vol. II., chap. XIII ; Nägeli, "Die Stärkekörner," Zürich, 1858, 4°, Flückiger und Hanbury's "Pharmacographia."

INDEX.

INDEX.

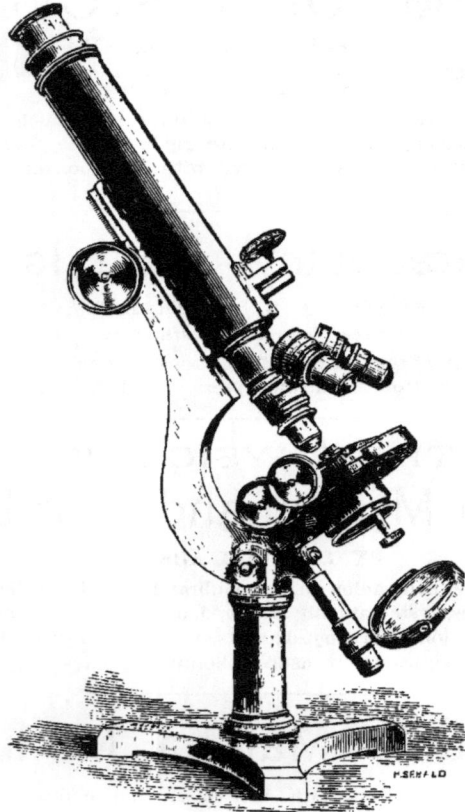

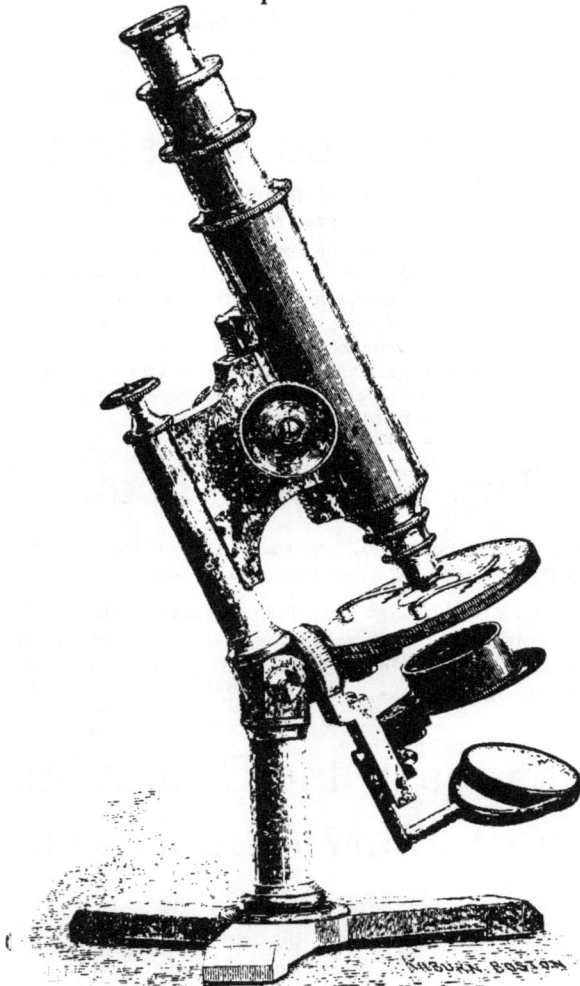